Water Science Reviews 2

T0296379

Water Science Reviews 2

Crystalline Hydrates

EDITED BY

FELIX FRANKS

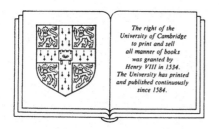

The right of the
University of Cambridge
to print and sell
all manner of books
was granted by
Henry VIII in 1534.
The University has printed
and published continuously
since 1584.

CAMBRIDGE UNIVERSITY PRESS

CAMBRIDGE

LONDON   NEW YORK   NEW ROCHELLE

MELBOURNE   SYDNEY

CAMBRIDGE UNIVERSITY PRESS
Cambridge, New York, Melbourne, Madrid, Cape Town, Singapore, São Paulo, Delhi

Cambridge University Press
The Edinburgh Building, Cambridge CB2 8RU, UK

Published in the United States of America by Cambridge University Press, New York

www.cambridge.org
Information on this title: www.cambridge.org/9780521341981

First published 1986
This digitally printed version 2008

A catalogue record for this publication is available from the British Library

ISBN 978-0-521-34198-1 hardback
ISBN 978-0-521-09100-8 paperback

Contents

The Structure of Ice-Ih

W. F. KUHS AND M. S. LEHMANN

Institut Laue–Langevin, 156X 38042 Grenoble Cedex, France

1. Introduction

Frozen water, in all its forms ranging from snowflakes to icebergs, has intrigued scientists for many centuries. Already Kepler [1] had speculated on the hexagonal shape of snowflakes, wondering what were the forces that transformed round droplets of water into the beautiful stars that he saw in falling snow. Although he did not resort to an atomic picture, he used a concept with formation by the cold of very small, identical particles, which would grow from an octahedral origin. There would thus be three orthogonal growth directions, leading to the formation of six branches in the star, and a hexagon could then form by a flattening of the star along one of the three-fold axes. Kepler discussed the form of different natural objects, ranging from honeycombs over pomegranates to different crystal forms, and he ended by proposing that different fluids would have in-built abilities that would lead to different forms when frozen.

Since then scientists have been equipped with a multitude of tools that allow them to go well beyond visual observation and speculation in the study of nature, but water in all its forms has always been a subject of interest. In this paper we shall deal with one small domain, namely the structure of ice-Ih. Methods of structure analysis originated in the first quarter of this century. Again ice was among the first compounds to be looked at with X-rays, and when neutron scattering became viable a powder spectrum of ice was among the first to be recorded.

Over the years methods of measurement and interpretation have developed, and at regular intervals ice has been studied, leading to a more and more precise and detailed picture of the atomic distribution. We shall report on this, and we shall also go on to discuss diffraction measurements, which have aimed at clarifying the pattern of disorder in ice.

Structural information is not only concerned with the atomic locations, but also with the interatomic arrangement, and this can be inferred from spectroscopic methods. In this review we have therefore also tried to extract what can be learned about distances and motions from spectroscopy, and to what extent there is agreement between the data obtained by the different methods.

Finally, theory should not be forgotten, so apart from the interpretations of observations given when needed, we report on the progress made in understanding the crystal structure with the aid of theoretical computational chemical methods.

The structure of ice-Ih seems, at first sight, a very limited topic. However, the literature on the subject is large, so to keep within a reasonable size the selection has been rather restrictive. We hope nevertheless that although only indicating the tip of the iceberg, we succeed in conveying the many exciting aspects of hexagonal ice.

2. Crystallographic studies

Ice-Ih is one of the most thoroughly disordered crystalline materials. The high translational symmetry is not retained at the level of the crystallographic unit cell. The symmetry elements of the crystallographic space group act only on the time- and space-averaged atomic densities, and there is no doubt that in this sense the symmetry in ice-Ih is very high indeed. In crystallographic studies sampling is over the instantaneous configurations in time and space provided that the interaction time of the radiation is short compared to the atomic motion and provided that the coherence length of the radiation is shorter than the size of the coherently scattering sample. Both conditions are in general fulfilled for studies on ice-Ih single crystals or powders with neutron or X-ray radiation. However, the atomic configurations in one specific unit cell do not obey the symmetry governing the time and space average. Static and/or dynamic disorder will produce local configurations with much lower symmetry. It is difficult to deduce from the time- and space-averaged picture the local atomic arrangements. Unravelling atomic disorder very often requires independent information from techniques which can probe local arrangements. By combining spectroscopic (see section 4) and nuclear magnetic resonance (see section 5) with high-resolution diffraction data, a three-dimensional picture of the local atomic arrangements can be drawn.

Certainly the most useful crystallographic experiment on ice-Ih is performed with neutron radiation. Neutrons interact strongly with hydrogen, deuterium and oxygen atoms. The scattering power in the neutron case is a constant as a function of the scattering angle and similar for all three elements, whereas in the X-ray case it is proportional to the number of electrons and decreases with increasing scattering angle. The neutron scattering lengths [2] are $f_H = -3.74 \times 10^{-15}$ m, $f_D = 6.67 \times 10^{-15}$ m and $f_O = 5.81 \times 10^{-15}$ m. X-ray experiments are much less informative and provide useful data on the lattice constants and oxygen atoms only.

There is a slight complication in neutron scattering for atoms like hydrogen and deuterium, which have a nuclear spin. In this case the scattering depends on the relative orientation of the neutron spin and the spin

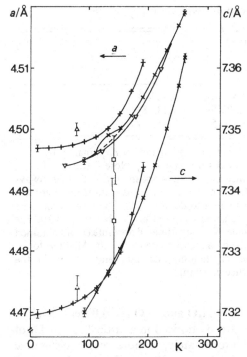

Figure 1. Lattice constants of ice-Ih: x, from La Placa and Post [4]; +, from Brill and Tippe [5]; and ▽, from Haltenorth [6]; all for H_2O. Single values are for D_2O: □, from Blackman and Lisgarten [7], and △, from Arnold *et al.* [8]. The lines are drawn as guides to the eye only. The error bars are plotted only at the lowest and highest temperature for each set.

of the nucleus. The orientation of the nuclear spin is different from atom to atom and thus also the interaction with neutrons. The spin incoherence gives rise to diffuse scattering, resulting in a considerably increased background intensity. This effect is weak in D_2O, but very strong in H_2O. However, in single-crystal neutral diffraction even the higher incoherent scattering cross-section of hydrogen is no serious problem, since the peak-to-background ratio for a Bragg peak is sufficiently high and since the incoherent scattering processes may be described as an absorption-like phenomenon, which is readily corrected for. Powder diffraction with its lower peak-to-background ratio suffers much from the strong incoherent scattering of hydrogen. Therefore neutron powder work is almost exclusively done on D_2O ices.

2.1. *The Lattice Constants*

The high resolution of X-ray powder diffractometers in principle allows the precise and accurate determination of the lattice constants in ice-Ih. Unfortunately there is still a considerable amount of disagreement about these basic quantities. The available data are plotted in figure 1. There are

Table 1. *Adopted† lattice constants of normal and heavy ice-Ih*

	H_2O (Å)		D_2O (Å)	
T(K)	a	c	a	c
60	4.4940(10)	7.318(2)	4.500$_5$(5)	7.329(8)
123	4.4962(10)	7.330(2)	4.503(5)	7.333(8)
223	4.5098(10)	7.347(2)	4.516(5)	7.351(8)

† The values for a_{H_2O} are taken from Haltenorth [6] and for c_{H_2O} mainly from La Placa & Post [4]. Note that Haltenorth does not give c_{H_2O}. Note further that a_{H_2O} agrees quite well in both references. Since in powder work the ratio of c/a is always more accurate than the individual lattice constants, we use the ratio from references La Placa and Post [4] and Brill and Tippe [5] to deduce c_{H_2O}. The lattice constants of heavy ice were deduced by combining some scattered single observations and the isotopic difference established in Lonsdale [3]. The estimated absolute errors obtained from Haltenorth [6] and the scatter in the data are quoted in brackets. All interatomic distances and angles calculated from crystallographic data and quoted in the following tables are based on these lattice constants.

indications that the lattice constants of H_2O and D_2O are different. Slightly larger volumes for heavy ice were already noticed in a critical review [3] of the early crystallographic data. Unfortunately, a more recent systematic study of lattice constants in heavy ice has not been performed. In normal ice-Ih anomalies along the *a*-direction were found near 120 K [4]. However, a re-examination [5] did not confirm this feature. Probably the most accurate study on the *a* lattice constant in H_2O was never published [6]. In general the internal agreement is rather poor by today's standards, and to obtain good estimates for D_2O is almost impossible in view of the widely scattered data points obtained at varying temperatures. For want of better measurements and in the balance of evidence, the lattice constants given in table 1 have been adopted in the following discussion. One should mention that the experimental scatter could be due to some real effect, e.g. inclusion of impurities, the size of crystallites, or even ageing effects, repeatedly invoked to explain discrepancies in other bulk measurements on ice-Ih. (See, for example [9], [10].)

2.2. The Atomic Probability Density

Diffraction data, or more accurately the intensities of the Bragg reflections observed at the nodes of the reciprocal lattice, carry information on the geometric arrangement of the atoms in a crystal. The intensity of a Bragg reflection, I_B, is proportional to the norm square, $F(Q) \cdot F(Q)^*$, of the structure amplitude $F(Q)$. The structure amplitude is given as

$$F(Q) = \sum_n f_n \exp{(\mathrm{i}\mathbf{r}_n \cdot Q)} \cdot W(Q) \qquad (2.1)$$

where $Q = 2\pi(ha^* + kb^* + lc^*)$ with h, k, l the Miller indices and a^*, b^*, c^* the reciprocal lattice constants. The position of atom n is described by $r_n = xa + yb + zc$ with x, y and z the fractional atomic coordinates and a, b, c the lattice constants. The individual atomic scattering power is dependent on the atomic scattering factor f_n. Atomic thermal motion reduces the measured intensities. In the harmonic approximation the weakening is given by the harmonic temperature factor

$$W(Q)_{\text{har}} = \exp\left[(2\pi)^{-2}(Q^{\text{T}} \cdot \beta \cdot Q)\right] \qquad (2.2)$$

with the symmetric tensor β describing the anisotropy of the mean-square displacements and Q^{T} the transpose of the scattering vector Q. Anharmonicity of the thermal motions or unresolved atomic disorder may occur and necessitate a more general temperature factor expression. [11] One very powerful generalized expression is based on the quasi-momentum expansion of the harmonic temperature factor

$$W(Q)_{gnrl} = W(Q)_{\text{har}} \cdot (1 - \tfrac{1}{6}i\gamma \cdot Q^3 + \tfrac{1}{24}\delta \cdot Q^4 + \ldots) \qquad (2.3)$$

where γ and δ are the quasi-moment tensors of ranks 3 and 4 respectively.

The atomic positions, the harmonic and anharmonic thermal parameters are determined from the intensities by a least-squares procedure. The resulting atomic probability density function (*PDF*) centred at the refined atomic positions is calculated from the derived parameters as

$$PDF(u)_{\text{har}} = \text{Det } P^{\frac{1}{2}}\left[(2\pi)^{-\frac{3}{2}} \exp\left(-\tfrac{1}{2}u^{\text{T}} \cdot P \cdot u\right)\right] \qquad (2.4)$$

with $P = 2\pi^2\beta^{-1}$, where β is the anisotropic mean-square-displacement tensor; u and u^{T} are the displacement vector and its transpose, respectively.

In the generalized form this trivariate *PDF* reads

$$PDF(u)_{gnrl} = PDF(u)_{\text{har}} \cdot (1 + \frac{1}{3!}\gamma \cdot H_3(u) + \frac{1}{4!}\delta \cdot H_4(u) + \ldots) \qquad (2.5)$$

with H_3 and H_4 being the Hermite polynomials of order 3 and 4. The higher order terms in brackets describe the anharmonic deformations of the harmonic (Gaussian) probability density. The calculated anharmonic deformation densities have been found to be very useful in detecting weak atomic disorder [11].

For an atom in a perfectly harmonic potential the mean and equilibrium positions are identical, while they are different in general in the case of anharmonicity. Employing the harmonic description (equations (2.1) and (2.2)) for anharmonic atomic potentials is certainly not adequate and corrections have to be applied to obtain the equilibrium positions (see equation (4.32)). By employing the generalized expressions (equations (2.3) and (2.5)) both equilibrium and mean positions are obtained. The accuracy of their determination is only limited by the adequacy of the chosen mathematical description, and the quasi-momentum expansion underlying equations (2.3) and (2.5) has proven to be adequate even in the case of strong anharmonicities [11]. Moreover, the generalized expressions describe not only anharmonic modifications of atomic densities, but also their curvilinear shape due to librational motions. Again corrections have to be made when harmonic models are used. In the case of librations of hydrogen or deuterium it is generally assumed that they 'ride' on the oxygen atom. The riding correction is given by [12]

$$\Delta r_{H(D)} = r_{O-H(D)} \langle \phi^2 \rangle \tag{2.6}$$

where $\langle \phi^2 \rangle$ is the mean-square angular displacement of librational motion. The best estimate of $\langle \phi^2 \rangle$ in disordered systems like ice-Ih is obtained from spectroscopy (see section 4).

2.3. The Averaged Structure of Ice-Ih

The time- and space-averaged structure of ice-Ih can be obtained with high precision and accuracy from neutron diffraction experiments, preferably on single crystals. The symmetry and unit cell of the averaged structures were correctly described by Barnes [13] from single-crystal X-ray data. Ice-Ih crystallizes in the hexagonal space group $P6_3/mmc$ with oxygen atoms placed at the four-fold position [14] (f): $\pm(\frac{1}{3}, \frac{2}{3}, z; \frac{2}{3}, \frac{1}{3}+z)$. The oxygen framework forms a very open structure and is shown in figure 2. The positions of the hydrogen atoms were first established by Wollan et al. [15] from neutron power data. Hydrogen or deuterium atoms are distributed over two sites, one four-fold position (f) and one twelve-fold position (k) [14]: $\pm(x, 2x, z; x, \bar{x}, z;$ $2\bar{x}, \bar{x}, z; x, 2x, \frac{1}{2}-z; x, \bar{x}, \frac{1}{2}-z; 2\bar{x}, \bar{x}, \frac{1}{2}-z)$. The averaged occupancy on these sites is 50% in agreement with Pauling's [16] half-hydrogen model. The atomic arrangement is shown in figure 3.

In a single-crystal neutron diffraction study performed at 123 and 223 K, Peterson and Levy [17] have provided a full description of the averaged structure of ice-Ih. The anisotropic atomic mean-square displacements were obtained and analysed. Pauling's half-hydrogen model was fully confirmed and the mean-square displacements of deuterium relative to oxygen essentially along the bond directions were found to be in agreement with spectroscopy. Later Chidambaram [18] proposed a bent hydrogen-bond model; displacing the hydrogen atoms from the c-axis and thus allowing the water angle to be preserved in ice-Ih, gave as good an agreement with Peterson and Levy's [17] data as the simple half-hydrogen model. Chamberlain et al. [19] studied doped H_2O ice at 77 K and essentially confirmed the validity of the half-hydrogen model for H_2O at low temperatures. Recently Kuhs and Lehmann [20–23] performed a series of high-resolution neutron diffraction studies on H_2O and D_2O ice-Ih in order to clarify some questions concerned with the molecular geometry. Doubts were raised (for example [24]) as to the correctness of the long O–H bond found in the earlier crystallographic work

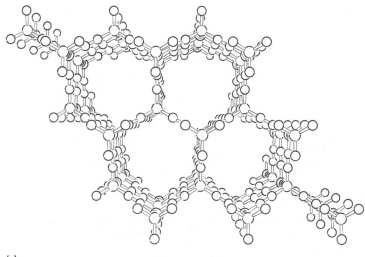

(a)

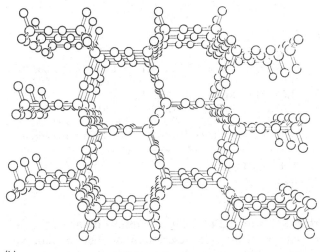

(b)

Figure 2. Framework of the oxygen atoms viewed along the c-axis (a) and along the a-axis (b) showing the open structure of ice-Ih.

[17, 19] and the validity of the bent hydrogen-bond model [18] was awaiting experimental proof. Data on H_2O and D_2O ice-Ih were collected at 60, 123 and 223 K and corrections for absorption (H_2O only) and first-order thermal diffuse scattering applied. Extinction was corrected in a first structure refinement stage. After averaging of equivalent reflections, the final structure

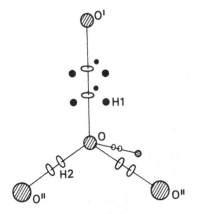

Figure 3. Averaged atomic arrangement in one oxygen tetrahedron of ice-Ih. The hydrogen positions are half occupied ('half-hydrogen model'). The bent hydrogen model [18] is indicated as a further splitting perpendicular to the *c*-axis.

Table 2. *Averaged molecular geometry in ice-Ih*

	H_2O			D_2O		
	60 K†	123 K†	223 K†	60 K†	123 K†	223 K‡
O–H1	1.005(2)	1.009(2)	1.000(4)	1.003(1)	1.002(1)	1.000(9)
O–H2	1.005(1)	1.005(1)	0.997(2)	$0.999_5(1)$	1.001(1)	1.007(5)
H1–H2	1.638(3)	1.643(3)	1.632(5)	1.633(1)	1.634(1)	1.635(12)
H2–H2	1.645(2)	1.641(2)	1.628(4)	1.634(2)	1.636(1)	1.648(9)
H1–O–H2	109.15(13)	109.42(14)	109.52(24)	109.31(6)	109.30(6)	109.10(30)
H2–O–H2	109.79(13)	109.53(14)	109.43(24)	109.63(6)	109.65(6)	109.86(26)

† Data from Kuhs and Lehmann [23].
‡ Data from Peterson and Levy [17].

refinements were performed. Tables 2 to 4 give the results obtained for the half-hydrogen model with an harmonic description of thermal motions (equations (2.2) and (2.4)). Although oxygen disorder and bent hydrogen bonds have not been included at this stage, valuable information is gained from this very global averaging over all local configurations. Several points emerge from an intercomparison of the results as a function of temperature and composition. As expected, the intermolecular distances increase in H_2O and D_2O as a function of temperature, while the change in the averaged intramolecular geometry is barely significant. The slight differences in bond length in H_2O and D_2O are due to the different magnitudes of stretch anharmonicity and libration. The intermolecular distances along and oblique

Table 3. *Averaged hydrogen-bond geometry in ice-Ih*

	H₂O			D₂O		
	60 K†	123 K†	223 K†	60 K†	123 K†	223 K‡
O–O'	2.748(1)	2.752(1)	2.760(1)	$2.754_5(1)$	2.755(1)	2.752(8)
O–O"	2.749(1)	2.752(1)	2.759(2)	2.753(1)	2.755(1)	2.765(1)
O–H1	1.743(2)	1.743(1)	1.758(3)	1.752(1)	1.753(1)	1.751(10)
O–H2	1.744(1)	1.747(1)	1.763(2)	1.754(1)	1.754(1)	1.759(6)
O–H1–O'	180.00	180.00	180.00	180.00	180.00	180.00
O–H2–O"	179.73(20)	179.93(21)	179.76(37)	179.99(9)	179.96(10)	176.8(2)
O'–O–O"	109.32(2)	109.37(2)	109.36(4)	109.30(1)	109.32(1)	109.55(15)
O"–O–O"	109.62(2)	109.57(2)	109.58(4)	109.64(1)	109.62(1)	109.40(2)

† Data from Kuhs and Lehmann [23].
‡ Data from Peterson and Levy [17].

Table 4. *Total† atomic mean-square displacements in ice-Ih*

	H₂O			D₂O		
	60 K‡	123 K‡	223 K‡	60 K‡	123 K‡	223 K§
O//c	0.01466(14)	0.02467(18)	0.04415(50)	0.01352(9)	0.02353(12)	0.02900(215)
O⊥c	0.01465(8)	0.02486(11)	0.04531(28)	0.01399(5)	0.02354(9)	0.03027(279)
H1//c	0.01904(49)	0.02812(58)	0.04906(175)	0.01605(20)	0.02495(22)	0.03762(355)
H1⊥c	0.03239(35)	0.04171(42)	0.05828(80)	0.02593(14)	0.03457(16)	0.04091(279)
H2//O–O″	0.01913(34)	0.02870(45)	0.04767(134)	0.01662(13)	0.02570(15)	0.03356(320)
H2//a	0.03262(48)	0.04161(57)	0.05830(103)	0.02580(17)	0.03431(19)	0.04889(576)
H2⊥a	0.03234(48)	0.04125(55)	0.05893(115)	0.02568(16)	0.03409(19)	0.04040(424)

† The mean-square displacements quoted include possible disorder displacements.
‡ Data from Kuhs and Lehmann [23].
§ Data on D₂O at 223 K are taken from Peterson and Levy [17]. Since a correction for thermal diffuse scattering was not made, the values quoted are considerably too low and cannot be used for a more detailed analysis.

to the crystallographic c-axis are identical within the limits of error. However, the O–O–O angles are larger when two of the oxygens are in the hexagonal plane, leading to a slight compression along the c-axis. The c/a ratio is therefore smaller than the ideal value of 1.633. Likewise, at least at low temperatures, the H1–O–H2 bond angle is smaller than the H2–O–H2 angle. This means that in the global average the hydrogen bond is very close to linear. The thermal motions of the oxygen atoms are fairly isotropic and slightly smaller in D_2O, as expected. Anisotropy in the atomic motion is pronounced for hydrogen and deuterium for directions parallel and perpendicular to the bond, while deviations from the isotropic motion in the plane perpendicular to the bond are insignificant. The two independent hydrogen positions are identical in their thermal motions. A bent hydrogen-bond model as proposed by Chidambaram [18] is not in agreement with these observations. An additional split of the H1 position will produce a significantly increased thermal motion perpendicular to the bond compared to H2. However, the simultaneous presence of bent hydrogen bonds in the two symmetrically independent directions is quite possible and in fact confirmed by further analysis.

2.4. *Molecular and Hydrogen-Bond Geometry in Ice-Ih*

In a disordered system it is not possible to deduce the local geometry from crystallographic data alone. However, diffraction data, measured to a resolution commensurate with the disorder displacement, can give valuable information about the spatial atomic disorder. In the structure of ice-Ih the disorder components for the two hydrogens are fairly large ($\simeq 0.75$ Å), but small (a few hundredths of an angström) for a possible bent hydrogen bond or possible oxygen disorder. The first case is well resolved, whereas in the other cases the disorder is only a fraction of the thermal root-mean-square displacements and therefore impossible to resolve. However, the mean-square disorder displacements cause deviations from a simple harmonic description of thermal motions and may be detected with high-resolution diffraction data. Such deformations are visualized in the anharmonic deformation density function calculated with equation (2.5). Another way of unravelling weak atomic disorder is the intercomparison of crystallographic and spectroscopic mean-square displacements. The spectroscopic values do not include disorder components, while the crystallographic values do. Hence, the disorder component of oxygen along the O–H(D) bond is obtained from

$$\langle u^2 \rangle_{O, dis\|} = \langle u^2 \rangle_{O\|} - \langle u^2 \rangle_{H(D)\|} + \langle u^2 \rangle_{H(D), spec\|} \qquad (2.7)$$

where $\langle u^2 \rangle_{O\|}$ and $\langle u^2 \rangle_{H(D)\|}$ are the crystallographic mean-square displacements of oxygen and hydrogen (deuterium) and $\langle u^2 \rangle_{H(D), spec\|}$ is the spectroscopic mean-square displacement of the O–H(D) stretch (see table 6 in section 4), all calculated parallel to the O–H(D) bond. Here and in the following, the

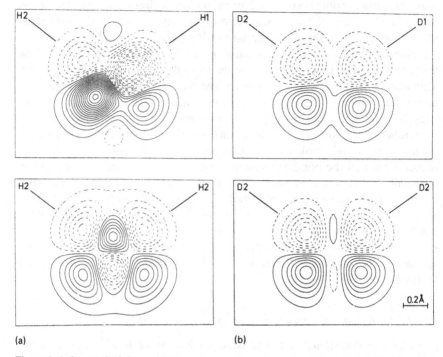

(a)   (b)

Figure 4. Anharmonic deformation density map of normal and heavy ice-Ih at 123 K in the planes of the idealized water molecules. The deformations shown are due to third-order terms in equation (2.5) and give the deviations from the harmonic (Gaussian) probability density.

average of the two independent hydrogen atoms is used; this procedure is justified by the fact that the difference in their mean-square displacement is insignificant. The use of equation (2.7) tacitly assumes that there is no hydrogen disorder parallel to the O–H(D) bond further than the distribution over the two half-hydrogen sites, which is a very reasonable assumption. Disorder of the molecule as a whole is of course not included in $\langle u^2 \rangle_{O, \text{dis}\parallel}$. Similarly, the disorder component of hydrogen (deuterium) perpendicular to the O–H(D) bond (bent disorder) is obtained from

$$\langle u^2 \rangle_{H(D), \text{dis}\perp} = \langle u^2 \rangle_{H(D)\perp} - \langle u^2 \rangle_{O\perp} - \langle u^2 \rangle_{O, \text{dis}\perp} - \langle u^2 \rangle_{H(D), \text{spec}\perp} \qquad (2.8)$$

where $\langle u^2 \rangle_{H(D)\perp}$ and $\langle u^2 \rangle_{O\perp}$ are obtained from crystallography and $\langle u^2 \rangle_{H(D), \text{spec}\perp}$ is assumed to originate in the librational motion only (see table 9 in section 4). The oxygen disorder perpendicular to the bond $\langle u^2 \rangle_{O, \text{dis}\perp}$ is assumed to be identical to $\langle u^2 \rangle_{O, \text{dis}\parallel}$. The interpretation of the deduced disorder displacements is not straightforward. However, combined with the anharmonic deformation maps some insight into the spatial atomic distribution may be obtained. The interpretation in terms of molecular or hydrogen-bond geometry is even more ambitious. Assumptions about the

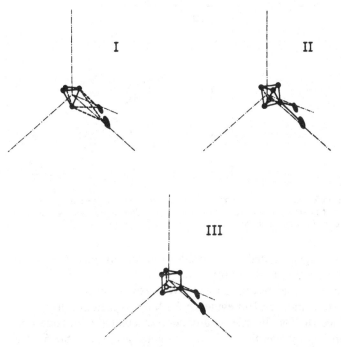

Figure 5. Three possible models for the oxygen disorder in ice-Ih resulting in tetrahedral, octahedral or trigonal space averaged displacement pattern.

approximate molecular geometry have to be made to infer these local geometries from the vast number of possible configurations in the averaged structure.

The anharmonic deformation density of the oxygen position in D_2O and H_2O ice-Ih at 123 K are shown in figure 4 [22]. These deformations are due to highly significant third-order terms in the expansion given in equation (2.5). Regions of positive density are located in directions opposite to the O–H(D) bond and to a lesser extent in the bisector of the water molecule. Since there is no reason to assume strong anharmonicities in the thermal motions of oxygen – the anharmonic effects in ice are dominated by hydrogen-bond bending [25] and not by O...O stretching [26] – the deformation must be due to oxygen disorder. Kuhs and Lehmann [22] proposed three disorder models, shown in figure 5, all consistent with the observed anharmonic deformation densities. Likewise, the anharmonic deformations in the hydrogen positions are shown in figure 6 in a section perpendicular to the hydrogen bond. The peakedness of the map is interpreted as an indicator for disorder [23]. However, the corresponding fourth-order terms are not highly significant and the experimental evidence for preferential disorder displacements manifested in third-order terms is even weaker. All anharmonic com-

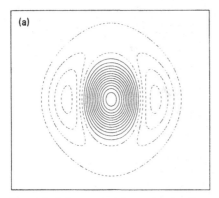 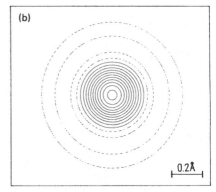

Figure 6. Anharmonic deformation density map of the deuteron in heavy ice-Ih at 123 K in a plane perpendicular to the c-axis: (a) for D2 and (b) for D1. The deformations shown are due to fourth-order terms in equation (2.5).

ponents in the hydrogen probability densities are largely masked by the pronounced harmonic thermal motions and only data from even lower temperatures to even higher resolution will give unequivocal results.

Fortunately there is independent evidence for both oxygen and hydrogen bond disorder in ice-Ih. The disorder components calculated with equations (2.7) and (2.8) are highly significant. The results are given in table 5. It is worthwhile to note that the correction of thermal diffuse scattering affects the magnitude of calculated disorder, underestimating the oxygen disorder and overestimating the bent disorder. It is possibly the neglect of this correction in Peterson's and Levy's work [17], which accidentally gave a reasonable agreement between crystallographic and spectroscopic O–D stretch displacements. The errors on the crystallographically-derived mean-square displacements are always much larger compared with spectroscopy and this has obscured any definite statement on this subject in the past. Chamberlain *et al.* [19] found hydrogen displacements perpendicular to the O–H bond which were larger than those calculated from spectroscopy, but in view of the experimental errors involved the agreement was found to be reasonable. The same argument was stressed by Kuhs and Lehmann [20], where in addition an incorrect reduced mass was used. Only the precision (cf. table 4) and accuracy of recent data allow one to establish these disorder components unequivocally [22, 23]. The interpretation in the case of the O–H(D) bond bending is straightforward; the root-mean-square displacement calculated from equation (2.8) gives the required disorder displacement. This is not so in the case of oxygen disorder. The oxygen displacement depends on the actual spatial disorder configuration. In general it will be larger than that calculated from equation (2.7), which represents the averaged displacement along the O–H(D) bond only. On the basis of the models shown in figure 5 the displacement of oxygen from its position in the idealized ice

Table 5. *Disorder mean-square displacements in ice-Ih*

	H_2O			D_2O	
	60 K	123 K	223 K	60 K	123 K
$\langle u^2 \rangle_{O\parallel}$	0.0007(4)	0.0016(6)	0.0025(16)	0.0009(2)	0.0015(2)
$\langle u^2 \rangle_{H\perp}$	0.0064(5)	0.0084(6)	0.0161(11)	0.0057(2)	0.0093(3)
$\langle u^2 \rangle_{mol}$	0.0028(4)	0.0019(4)	0.0041(4)	0.0021(2)	0.0013(2)

† $\langle u^2 \rangle_{O\parallel}$ is evaluated with equation (2.7), $\langle u^2 \rangle_{H\perp}$ with equation (2.8). $\langle u^2 \rangle_{mol}$ is obtained from the difference between the crystallographic mean-square displacements after subtraction of atomic disorder and the mean-square displacements obtained from specific heat data [27] given in table 10. The errors quoted are based on the experimental uncertainties of the crystallographic data; the precision of the spectroscopic entries in general is much higher than those errors. However, in the case of $\langle u^2 \rangle_{H\perp}$ the accuracy of the spectroscopic estimate for $\langle u^2 \rangle_{H(D),\,spec\perp}$ in equation (2.8) is probably low, since the effect of inter- and intramolecular bending has not been taken into account. All errors quoted are based on the crystallographically estimated errors solely and give only lower limits of the absolute error.

structure was estimated to be 0.06(2) Å. A further confirmation of this number comes from structure refinements of the three proposed disorder models for D_2O at 123 K [22]. With O–D stretch displacements constrained to the spectroscopic values, the above-mentioned disorder displacements agree within the limits of error. For the marginally preferred model 2 [22, 23] with oxygen displaced along the bisectors, a displacement of 0.05(2) Å was obtained. From these refinements one could in principle derive the spread in intermolecular distances and compare it with the spectroscopic estimate. However, the models used are grossly oversimplified, since they completely ignore the influence of the neighbouring water molecules. Therefore only crude estimates can be obtained; the spread is a few hundredths of an ångström with some differences for the different models [23]. The calculated mean intermolecular distance in all disorder models is shifted by several thousandths of an ångström to slightly larger values compared with the values of the original half-hydrogen model. However, the crudeness of the refined models does not allow any specific numbers to be quoted. Similar arguments hold for the mean value and spread of the hydrogen-bond distance $r_{H(D)\dots O}$ with the only difference that the shift definitely is a few hundredths of an ångström upwards from 1.75 to roughly 1.8 Å, in good agreement with spectroscopic estimates (see section 4). The hydrogen-bond angles are very probably bent by a few degrees with some spread, which at present cannot be quantified further. The influence of the nearest neighbours on the actual hydrogen-bond geometry is manifest in the molecular disorder. The oxygen disorder established from equation (2.7) does not contain molecular disorder, in which oxygen and hydrogen atoms are simply translated from their averaged positions to accommodate a given nearest-neighbour configuration. The magnitude of this effect can be calculated from a comparison of mean-square displacements obtained from specific heat [27] and crystallography, given in table 10 in section 4. The molecular disorder is substantial in H_2O and D_2O ice, of similar magnitude to the oxygen disorder, and increases with temperature. This increase is expected to show up in an increased spread of intermolecular distance, and indeed is observed by spectroscopy (see section 4). Moreover, molecular disorder will change intermolecular distances in a way which cannot be separated from diffraction data.

The situation as regards the intramolecular geometry is slightly more favourable. Molecular disorder is of no importance and estimates of oxygen disorder and hydrogen-bond bending are available (see table 5). From the relevant force constants, the intramolecular distances are expected to increase only marginally with increasing temperature, while the temperature dependence of the intramolecular angle is less certain, but probably small. Depending on the models used, the O–H(D) bond lengths obtained at 123 K vary between 0.975 and 0.987 Å [22]. Model 2, which is slightly preferred in the refinement, gives 0.979(5) Å. The increase of oxygen disorder with increasing temperature is hardly significant (see table 5). Likewise the differences between H_2O and D_2O ice are insignificant. Attempts to calculate the intramolecular angle were made [23], but no unequivocal answer has been obtained. With oxygens displaced according to model 2 one obtains, with the hydrogen-bond-bending displacements given in table 5, intramolecular angles closer to the water angle of 104.45° than to the tetrahedral angle of 109.47°. The decrease in angle with increasing temperature is probably not real; rather it is caused by the neglect of intermolecular bending in the calculation of the relative displacements of oxygen and hydrogen (see section 4). Estimates of the intramolecular angle, however, may be obtained by combining the measured bond lengths with the intramolecular H–H distance r_{H-H} deduced from nuclear magnetic resonance data (see section 5). The crystallographic r_{H-H} (r_{D-D}) distances are subjected to the same uncertainties as the bond angle estimates discussed above. At low temperatures and within large errors they do however agree with the nuclear magnetic resonance (NMR) value of approximately 1.58 Å.

The details of the atomic arrangements in ice-Ih are still not known accurately. Three models explain the observed atomic probability densities, and the diffraction data do not allow a definite choice. All conclusions based on any one of these three models are therefore burdened with much uncertainty. Accuracy and precision of interatomic distances and angles quoted above are certainly lower than are usually met in crystallographic work on ordered structures. Nevertheless, there is strong evidence that the much debated long O–H(D) distance, repeatedly quoted in the crystallographic literature, is entirely due to oxygen disorder. The inclusion of oxygen disorder and O–H(D) bond bending brings crystallography back into agreement with spectroscopic and NMR data. The fine details of the structure, however, still remain obscure. To learn more about the three-dimensional atomic arrangements will demand data to a higher resolution measured at lower temperatures than those currently available.

3. Experimental Observations of the Ice Rules and Local Ordering

The most direct experimental information about the organization of three-dimensional crystal structures comes from diffraction methods, and is obtained by analysis of Bragg-reflection data as described above. The result is the atomic arrangement averaged over time and space. As the interaction time is short compared with the oscillation time of the atom, the average is over a series of instantaneous observations of the atomic distribution for a domain of the crystal covered by the coherence length of the radiation. This depends on the monochromatic nature of the beam and the crystallinity and is normally greater than 10^3 Å. A study of the Bragg reflections will therefore give no direct indication of the *local* arrangement, but only an average picture for a large number of unit cells.

3.1. *The Ice Rules*

The crystallographic studies reported above allow for a large number of possible local configurations, and other methods have therefore also been used to obtain the information needed. From Raman and infrared spectra Bernal and Fowler [28] deduced that molecules in water and ice did not differ from molecules in steam, except for small deformations, and this observation was followed up by Pauling [16] who suggested a series of rules governing the location of hydrogen atoms – the ice rules. They state:

(1) The water molecule in ice resembles the water molecule in the gas phase.

(2) Each water molecule is oriented so that its two hydrogen atoms are directed approximately toward two of the four oxygen atoms which surround it tetrahedrally, forming hydrogen bonds.

(3) Only one hydrogen atom is positioned between each neighbouring oxygen–oxygen pair.

(4) Under ordinary conditions the interaction of non-adjacent molecules is not such as to stabilize appreciably any one of the many configurations satisfying the preceding conditions with reference to the others.

These results were then used to estimate the entropy, S_0, of ice at very low temperatures. The entropy is given by

$$S_0 = k \ln W \tag{3.1}$$

where k is Boltzmann's constant, and W is the number of possible configurations of the ice lattice. Pauling estimated W in two different ways.

Using condition 2 it is obvious that the water molecule can take up six orientations. Moreover, of the four hydrogen bonds in which one water molecule participates, two are occupied by its hydrogen atoms and two are unoccupied. The chance that a given direction is available to a hydrogen atom is therefore $\frac{1}{2}$, and as there are two hydrogen atoms to be placed, the probability that this can be done is $\frac{1}{2} \times \frac{1}{2} = \frac{1}{4}$. If there is a total of N molecules, the total number of configurations is then $W = (6/4)^N = (3/2)^N$.

Another approach is first to neglect condition 1. In this case $W = 2^{2N}$, as each of the $2N$ hydrogen bonds can exist in two states. Similarly, there are $2^4 (= 16)$ possible configurations of hydrogen in the four bonds holding one oxygen atom. Of these only six are permitted by condition 1, and the total number of configurations is therefore $W = (6/16)^N \times 2^{2N} = (3/2)^N$.

We therefore arrive at an entropy of

$$S_0 = k \ln (3/2)^N = R \ln (3/2) = 3.38 \text{ J mol}^{-1} \text{deg}^{-1} \tag{3.2}$$

which is in good agreement with the observed value [29, 30] of $3.41 \text{ J mol}^{-1} \text{deg}^{-1}$.

3.1.1. Configurational Statistics. The derivation of the entropy explicitly assumes, according to condition 4, that none of the configurations is preferentially stabilized due to interactions between non-adjacent molecules. Moreover, it assumes that a given choice of local configuration has no influence on the possible configurations at neighbouring sites, and this fails because there are closed rings in the lattice. The smallest ring that can be formed is a six-membered ring, and the constraint on the configurations coming from this ring has been discussed by Hollins [31]. Figure 7 shows such a ring configuration.

Walking along the ring from 1 to 6 we encounter a set of signs for each atom, namely $- +, - -, + -, + +, - +, - +$ for atoms 1, 2, 3, 4, 5 and 6, respectively. Condition 3 requires that pairs of signs straddling commas (corresponding to the hydrogen bonds) must be of unlike sign, and that,

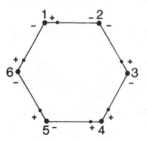

Figure 7. Six-membered ring in ice. + indicates that hydrogen is present; − that it is absent.

(a)

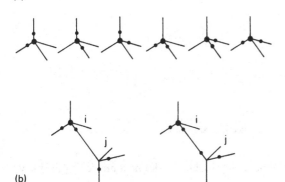

(b)

Figure 8. (a) The six allowed vertices in the ice lattice. The black dots represent the two hydrogens bound to the oxygen. (b) compatible and incompatible pairs of neighbouring vertices i and j.

likewise, at the closure from 6 to 1 the last sign of 6 and the first sign of 1 are opposite. There is no constraint on the individual pair of signs, but as the first and last sign must be opposite, there must be an even number of pairs (or zero pairs) within the molecules, which have like signs. Counting the probability under these constraints leads to a correction term to the above value of W, which now becomes

$$W = (3/2)^N (1+(1/729))^{2N}. \tag{3.3}$$

There are further interactions involving larger rings, and moreover the rings overlap partially. Better estimates of W including these have been done by Nagle [32, 33], who extended an expression first used for ice by Di Marzio and Stillinger [34], and we shall discuss this in some detail.

As mentioned above, there are six allowed configurations around one oxygen, a vertex, of the hexagonal ice lattice, and these are shown in figure 8(a). For two neighbouring vertices i and j a compatibility function is defined as

$$A(\xi_i, \xi_j) = \begin{cases} 1 & \text{if } \xi_i \text{ and } \xi_j \text{ are compatible} \\ 0 & \text{if } \xi_i \text{ and } \xi_j \text{ are incompatible} \end{cases} \tag{3.4}$$

where the state parameter ξ refers to one of the six configurations in figure 8(a). Figure 8(b) shows cases where $A(\xi_i, \xi_j)$ is 1 or 0, respectively. The total number of configurations in an ice crystal can now be written as

$$W = \sum_{\{\xi\}} \prod_{i<j} A(\xi_i, \xi_j) \tag{3.5}$$

where $\sum_{\{\xi\}}$ means to sum over the set of all 6^N different combinations of ξ

arrangements at each vertex i, $\prod_{i<j}$ is the product over nearest neighbours

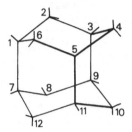

Figure 9. Part of ice lattice. The heavy connections show a graph in the lattice.

with each pair taken once. The next step is then to define a new compatibility function defined as

$$a(\xi_i, \xi_j) = \begin{cases} +1 & \text{if } A(\xi_i, \xi_j) = 1 \\ -1 & \text{if } A(\xi_i, \xi_j) = 0 \end{cases} \tag{3.6}$$

and W can then be rewritten

$$W = (3/2)^N \sum_{\{\xi\}} \prod_{i<j} (1/6)^{\frac{1}{2}} [1 + a(\xi_i, \xi_j)]. \tag{3.7}$$

Note that for a configuration where all the $2N$ bonds are compatible, the product of the $(1/6)^{\frac{1}{2}}$ factor becomes $(1/6)^N$ and the product of $[1 + a(\xi_i, \xi_j)]$ becomes 2^{2N}, giving altogether $(2/3)^N$.

The product can now be expanded as a series, where the zeroth term will contain no a, the first term will contain one a, the second term will contain two as etc. An example of a term in the product series is

$$a(\xi_3, \xi_4)\, a(\xi_4, \xi_5)\, a(\xi_5, \xi_6)\, a(\xi_5, \xi_{11})\, a(\xi_{10}, \xi_{11}) \tag{3.8}$$

which can be related to a graph in the ice lattice. This is shown in figure 9. The total contribution from this term is now obtained by summing over all the ξ configurations for the vertices involved, and by grouping these sums it is easily seen that many contributions are zero. This comes from the two identities

$$\sum_{\xi_i} a(\xi_i, \xi_j) = 0 \tag{3.9}$$

and

$$\sum_{\xi_i} a(\xi_i, \xi_j)\, a(\xi_i, \xi_k)\, a(\xi_i, \xi_l) = 0 \tag{3.10}$$

which are easily seen to hold by inspection of figure 8(a). This implies that any graph containing vertices with either one or three incident edges gives a zero contribution to the sum of products. The example given in figure 9 is such a case. Only closed graphs do therefore contribute, and the smallest such graph is the cycle including six oxygen atoms. In figure 9 such a cycle

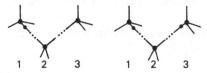

Figure 10. Graph with three vertices, 1, 2 and 3. Only the location of the hydrogen atoms relevant to the cycle are given. Vertex two has six possible configurations, and summing over these $a(\xi_1, \xi_2)\, a(\xi_2, \xi_3)$ is $+2$ for the left-hand example, and -2 for the right-hand example.

is given for example by vertices 1, 2, 3, 4, 5 and 6. The contributions of different kinds of cycles, the weights, now have to be considered. The simplest of these are cycles where all vertices have two incident edges and these are the only ones that we shall discuss here. It is easy to see from figures 8(a) and 10 that

$$\sum_{\xi_2} a(\xi_1, \xi_2)\, a(\xi_2, \xi_3) \tag{3.11}$$

is either $+2$ or -2. Proceeding by induction, it can then be shown that

$$\sum_{\xi_2 \dots \xi_{i-1}} a(\xi_1, \xi_2)\dots a(\xi_{i-1}, \xi_i) \tag{3.12}$$

is either $+2^{i-2}$ or -2^{i-2}. The plus sign occurs when one of ξ_1 and ξ_i contributes one hydrogen, while minus occurs when ξ_1 and ξ_i contribute two or no hydrogen atoms. Finally we can look at a ring with n vertices and set $i = n+1 = 1$. To complete the sum we must sum over ξ_1, which in four cases contributes one hydrogen atom to the cycle and in two cases none or two hydrogen atoms. The final sum therefore becomes

$$\sum_{\xi_1, \dots \xi_n} a(\xi_1, \xi_2)\dots a(\xi_n, \xi_1) = 4(2^{n-1}) - 2(2^{n-1}) = 2^n. \tag{3.13}$$

Including the factor $(1/6)^{\frac{1}{2}}$ n times then finally leads to a weight of $(1/3)^n$. This number must now be multiplied by the number of occurrences of a given cycle in order to obtain the total contribution to W. These values have been tabulated by Nagle [32] up to $n = 14$, and for example, for six-membered cycles there are $2N$ in all. Summing now over the various terms, and adding an estimate for the higher order terms Nagle [32] obtained

$$S_0 = h \ln (1.50687) = 3.410 \text{ J mol}^{-1} \text{ deg}^{-1} \tag{3.14}$$

in very good agreement with the experimental value.

3.2. *Diffuse Scattering*

Although the observations and calculations of the zero-point entropy thus support the ice rules, it is clearly of interest to obtain further confirmation, preferably on a three-dimensional structural level, and this can be done using diffuse scattering techniques.

The total elastic scattering per unit cell from a crystal within a unit of solid angle Ω can be expressed most generally as [35, 36, 37].

$$I(\Omega) = N^{-1} \sum_{ij} F_i F_j^* \exp\left[iQ(r_j - r_i)\right] \tag{3.15}$$

where the sum is over the molecules in the N unit cells, Q is the scattering vector ($|Q| = 4\pi \sin\theta/\lambda$), r_i is the vector to molecule i, and F_i is the form factor for molecule i, written as

$$F_i = \sum_k f_k \exp\left(-iQ \cdot r_k\right). \tag{3.16}$$

The sum is over the atoms in the molecule, f_k is the form factor for atom k, and r_k is the atomic position with respect to the origin defined by r_i.

Introducing now the different configurations ξ as discussed above, and defining probabilities $P(\xi_i)$ and $P(\xi_i, \xi_j)$, where $P(\xi_i)$ is the probability of finding molecule i in a given orientation, and $P(\xi_i, \xi_j)$ is the joint probability of i and j being oriented in given ways, $I(\Omega)$ can be written as a sum of three terms

$$I(\Omega) = I_B(\Omega) + I_D(\Omega) + I_C(\Omega). \tag{3.17}$$

The first term is the Bragg scattering and has been discussed in section 2 above. The second term gives rise to diffuse scattering which is not dependent on correlation, and has the form

$$I_D(\Omega) = \overline{F^2} - \overline{F}^2. \tag{3.18}$$

The two averages of F are defined as

$$\overline{F^2} = \sum_{\xi_i} P(\xi_i)\, F(\xi_i)\, F^*(\xi_i) \tag{3.19}$$

$$\overline{F} = \sum_{\xi_i} P(\xi_i)\, F(\xi_i). \tag{3.20}$$

The last term describes the correlation dependent diffuse scattering and is

$$I_C(\Omega) = N^{-1} \sum_{\xi_i} \sum_{\xi_j} \sum_{i \neq j} F(\xi_i) F^*(\xi_j)\, C(\xi_i, \xi_j) \exp\left[iQ(r_j - r_i)\right] \tag{3.21}$$

with $c(\xi_i, \xi_j)$ being the correlation function

$$C(\xi_i, \xi_j) = P(\xi_i, \xi_j) - P(\xi_i)\, P(\xi_j). \tag{3.22}$$

The experiment now consists of the measurement of the elastic scattering from a powder or a single crystal for scattering directions other than those defined by the Bragg condition, and then to compare this with a calculation of $I_C(\Omega)$ with $C(\xi_i, \xi_j)$ modelled in some manner. As the total scattering from a crystal contains both elastic and inelastic components, these must be separated experimentally. This is very difficult in the X-ray case [38], as the energy of the radiation, typically in the 10 to 50 keV region, is much larger than the meV change in energy that occurs when the radiation is scattered inelastically. Moreover the X-ray scattering cross-section, being proportional to the square of the number of electrons in the atom, is much smaller for hydrogen than for oxygen, thus giving a small signal for the atom of interest.

The situation is very different for neutron scattering. In this case the energy of the incident neutrons is comparable with the energy of the lattice vibrations, and interactions with these can therefore lead to relatively large changes in the energy of scattered neutrons. Moreover, the scattering by the different atoms is comparable, and it should be noted that it is necessary to use deuterium rather than hydrogen, as hydrogen has a large cross-section for incoherent scattering, which for the studies in question only adds to the general background of the measurement. As deuterium scatters better than oxygen, deuteration is not too serious, but it should of course be kept in mind when interpreting the results, and when extrapolations are made to normal ice-Ih.

Two methods of neutron energy analysis are available and have been used. One consists of time-of-flight techniques [39, 41], where a monochromatic pulsed beam is used. The energy analysis consists in the rejection of neutrons for which the flight time is different from that for an elastic scattering process, where the neutron does not change velocity at the sample. The other method involves the insertion in the scattered beam of an energy analyser [40] consisting of a crystal which only reflects the neutrons with a wavelength (and thus energy) corresponding to the value of the incident neutrons. The most recent experiments based on these two techniques have given energy resolutions which are in the range from 0.4 to 3.9 meV for time-of-flight measurements [39] using wavelengths of 5.12 and 2.56 Å, and 1.4 meV for a wavelength of 1.26 Å and crystal analyser techniques [40].

The very first determination of the diffuse scattering was performed as early as 1949 in connection with the first neutron diffraction study [15] of ice-Ih. This structure analysis was indeed one of the very first neutron diffraction studies ever undertaken. It was a powder measurement, and only the correlation-independent value $I_D(\Omega)$ could be estimated. Using measurements at small angles, where inelastic scattering is not important, the total diffuse scattering could be estimated. This contains both a small incoherent term from spin scattering, with a scattering cross-section $4\pi b_D{}^2$ (incoherent) of 2.4×10^{-24} cm², and the term $I_D(\Omega)$. For a two-site disorder $P(\xi_i)$ would be $\frac{1}{2}$, and $F(\xi_i)F^*(\xi_i)$ would be $4\pi b_D{}^2$ (coherent) = 5.60 b, so $I_C(\Omega) = \frac{1}{2}(4\pi b_D{}^2(\text{coherent})) = 2.80$ b. The sum of the two components was in agreement with the observed value, indicating thus the two-site disorder as predicted by the ice rules.

Further qualitative observations were made by Peterson and Levy [17], but

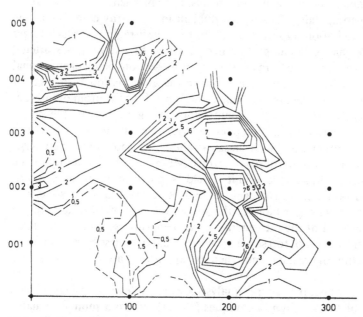

Figure 11. Elastic diffuse neutron scattering [39] (time-of-flight) of the $h0l$ plane of the ice-Ih at 123 K. The contour heights are indicated and are on relative scales.

only more recently have quantitative measurements been performed by Schneider [39], and Schneider and Zeyen [40]. Figure 11 shows a part of the reciprocal lattice studied by time-of-flight techniques.

This can now be compared with theoretical determinations of the term $I_C(\Omega)$, where the essential effort consists in expressing the correlation $C(\xi_i, \xi_j)$ between the possible ξ configurations at positions i and j of the lattice. The first calculation of this kind was reported by Villain and Schneider [42] using the random walk approximation (RWA). In this calculation the two possible positions of the hydrogen (deuterium) were described by a vector

$$r_{il} = R'_l + a_l S_{il} \tag{3.23}$$

where R'_l is the centre of the oxygen–oxygen line and $2a_l$ is a vector joining the two deuterium positions S_{il} then takes the values ± 1, and is treated like a pseudo-spin. $I_C(\Omega)$ can then be written (neglecting at this stage Debye–Waller factors)

$$I_C(\Omega) = 4b_D^2 \sum_{ij} \sin Q \cdot a_i \sin Q \cdot a_j \exp\left[iQ(R'_i - R'_j)\right] \chi_{ij}(Q) \tag{3.24}$$

where the first terms describe the geometry, and the last term is the Fourier transform of the correlation function

$$\chi_{ij}(Q) = \sum_{kl} \langle S_{ki} S_{lj} \rangle \exp [i\boldsymbol{Q}(\boldsymbol{R}_k - \boldsymbol{R}_l)] \tag{3.25}$$

where S_{ki} is the spin at point i in the cell at $\boldsymbol{R}_k$, and where the correlation function is of the type $\langle S_i S_j \rangle$ relating spins at two positions i and j in the lattice. The combination of S_i and S_j can now take four values $S_i^+ S_j^+$, $S_i^- S_j^-$, $S_i^+ S_j^-$ and $S_i^- S_j^+$ where $S^{\pm}$ corresponds to ± 1. If the probability of observing $S_i^+ S_j^+$ is called P_{ij}^{++}, that of $S_i^+ S_j^-$ is P_{ij}^{+-}, etc., then the probable value of $\langle S_i S_j \rangle$ is of the form

$$\langle S_i S_j \rangle \simeq P_{ij}^{++} - P_{ij}^{+-} \tag{3.26}$$

where P_{ij}^{++} and P_{ij}^{--} are assumed to be the same (parallel spin alignment) and $P_{ij}^{+-} = P_{ij}^{-+}$ (anti-parallel alignment). The probabilities are now estimated using the RWA. The edges between oxygen vertices are replaced by arrows following some rule, and a walker is allowed to walk along the arrows of the lattice. At each vertex he starts walking along one of the three adjoining bonds, and he is not allowed to go backwards. The walk from i to j can be done along many paths with different number of steps n, and $P_n^{++}(i/j)$ denotes the probability that having walked along the plus direction of i then after n steps the walker steps along j in the plus direction. $P_n^{+-}(i/j)$ and other terms are defined in a similar manner. The total probability is then, following the RWA, the sum over all possible ways, i.e.

$$\langle S_i, S_j \rangle = \sum_{n=-\infty}^{\infty} [P_n^{++}(i/j) - P^{+-}(i/j)]. \tag{3.27}$$

The ice rule 3 is ensured by the condition that the walker cannot step backwards, so that in principle he will only walk through every bond once. The condition is expressed in the values chosen for P_1s; $P_1^{+-}(i,i)$, for example, is zero. Violations of the rule can occur for paths longer than six steps, which is the smallest ring, but the error introduced was estimated to be less than 1%. Ice rule 1, which requires each oxygen to be bound to two hydrogens is fulfilled according to conservation laws for the arrow lattice, which gives that each vertex has the same number of incoming and outgoing arrows.

By a Fourier transformation the correlation function $\chi_{ij}(Q)$, $I_C(\Omega)$ can be calculated and the result for the plane $h0l$ is given in figure 12. There is good agreement between the location of observed and calculated maxima [39]. Even better agreement was obtained later by crystal analyser techniques [40], and a comparison between the observed and calculated diffuse scattering is shown in figure 13. The agreement is near quantitative and verifies the RWA approximation.

Another approach to the estimate of $I_C(\Omega)$ has been developed by Descamps and Coulon [36] and is based on principles similar to the graph

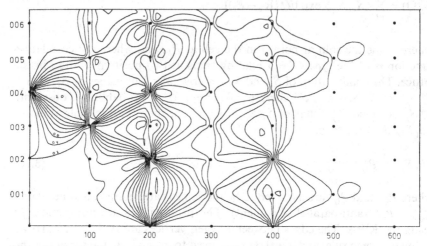

Figure 12. Elastic diffuse scattering [42] estimated by the RWA in the plane $h0l$.

expansion discussed above [32, 33]. In this case the expression for $C(\xi_i, \xi_j)$ (expression (3.21)) is used to rewrite $I_C(\Omega)$ to get

$$I_C(\Omega) = -N^{-1} \sum_{\xi_1} P(\xi_1) F(\xi_1) \sum_{\xi_j} \sum_{j>1} P(\xi_j) F^*(\xi_j) \exp[i\mathbf{Q}(\mathbf{r}_j - \mathbf{r}_1)]$$

$$+N^{-1} \sum_{\xi_1} \sum_{\xi_j} \sum_{j>1} P(\xi_1, \xi_j) F(\xi_1) F^*(\xi_j) \exp[i\mathbf{Q}(\mathbf{r}_j - \mathbf{r}_1)]$$

$$= -N^{-1} \bar{F}_1 \sum_{j>1} \bar{F}_j \exp[i\mathbf{Q}(\mathbf{r}_j - \mathbf{r}_1)]$$

$$+N^{-1} \sum_{\xi_1} \dots \sum_{\xi_N} P(\xi_1, \dots, \xi_N) \{ \sum_{j>1} F(\xi_1) F^*(\xi_j) \exp[i\mathbf{Q}(\mathbf{r}_j - \mathbf{r}_1)] \}. \quad (3.28)$$

To simplify the expression $I_C(\Omega)$ is only given for one molecule out of the four in the unit cell, and further reduction is obtained by developing the correlation terms only with respect to one fixed molecule.

The first term of the expression is easily dealt with, as $P(\xi_i)$ is known. For the second term the joint probability is rewritten in a manner similar to above as

$$P(\xi_1 \dots \xi_N) = (3/2)^2 / W \prod_{ij}^{2N} (1/6)^{\frac{1}{2}} [1 - C_{ij}(\xi_i) C_{ji}(\xi_j)] \quad (3.29)$$

where the product of the two Cs is taken only for the $2N$ neighbouring pairs of oxygen atoms. The definition of $C_{ij}(\xi_i)$ is now that

$$C_{ij}(\xi_i) = 1 \quad (3.30)$$

if the proton lying on the ij bond is near to i. Elsewhere $C_{ij}(\xi_i)$ is -1.

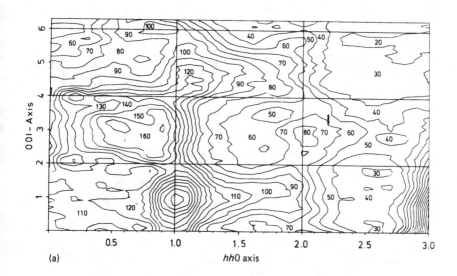

(a)

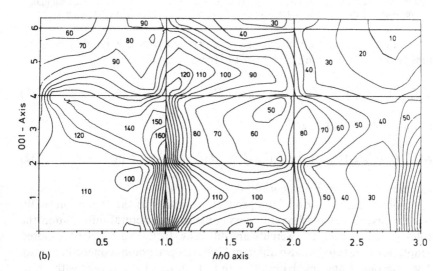

(b)

Figure 13. Elastic diffuse neutron scattering [40] (crystal analyser) in the plane *hhl* at 120 K. (a) is the observed diffuse scattering, while (b) is the theoretical RWA map, where a contribution from incoherent scattering has been added. Additional diffuse *l*-streaks centred on 110 and 330 have been added in (b).

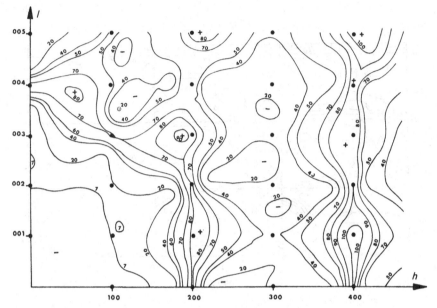

Figure 14. Elastic diffuse scattering as estimated using the graph method [36] in the $h0l$ plane. Debye–Waller factors corresponding to 123 K have been included in the calculation.

As for the calculation of entropy, the product is now expanded, and the first two terms involving C_{ij} are of the form

$$\Sigma\, C_{ij}(\xi_i)\, C_{ji}(\xi_j)\, \{\, \sum_{k\,>\,1}\, F(\xi_i)\, F^*(\xi_k)\, \exp\,[i Q(r_k - r_1)]\} \tag{3.31}$$

summed over edges ij, and

$$\Sigma\Sigma C_{ij}(\xi_i)\, C_{ji}(\xi_j)\, C_{pm}(\xi_p)\, C_{mp}(\xi_m)\, \{\, \sum_{k\,>\,1}\, F(\xi_1)\, F^*(\xi_k)\, \exp\,[i Q(r_k - r_1)]\} \tag{3.32}$$

summed over edges ij and pm.

As observed, the form of the products is similar to the expansion of W, but because of the additional factors containing structural information the rules for calculating the weights are different. In this case not only closed graphs are non-zero but also different kinds of open graphs connecting group 1 with group k and involving $F(\xi_1)$ and $F(\xi_k)$. In all the groups with up to eight edges were investigated and it was found that this generally led to rapid convergence, although the evolution depended on the Q value. The theoretical map corresponding to the observation depicted in figure 11 is shown in figure 14. Again there is good general agreement.

Altogether, therefore, the diffuse scattering and the theoretical estimates confirm the ice rules. It should be noted, though, that the most recent measurements [40] indicated some streaks along l through the 110 and 330

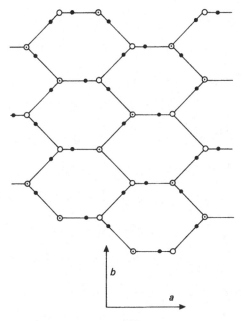

Figure 15. Section through the water layer found in the *ab* plane of copper formate tetrahydrate. Open circles represent oxygen; dots, hydrogen atoms. Out-of-plane hydrogen atoms are drawn with the out-of-plane hydrogen inside the circle representing the oxygen atom. In the disordered phase all in-plane hydrogen atoms are disordered.

reciprocal lattice points. This could indicate correlation or stacking-fault effects, but has not yet been the subject of further investigations.

3.3. *Two-dimensional Ice*

Because of the good general agreement between the observations and the calculated scattering, there is no strong indication of any order beyond that imposed by the ice rules. As pointed out by Stillinger and Cotter [43], this is quite remarkable, because interactions between neighbours alone would favour an anti-parallel arrangement of water molecule dipoles over a parallel arrangement. A possible explanation can be found in the studies of a two-dimensional layer structure of water, in copper formate tetrahydrate, which is believed to follow the ice rules. The hydrogen atoms in the layer are disordered above 234 K, and ordered below this temperature in a way indicated in figure 15. In this state, individual layers of water are ordered ferroelectrically with polarization along the *b*-axis. This polarization changes sign from plane to plane, thus producing in total an anti-ferroelectric ordering of the three-dimensional structure. The phase transition was studied from the elastic diffuse neutron scattering above the phase transition

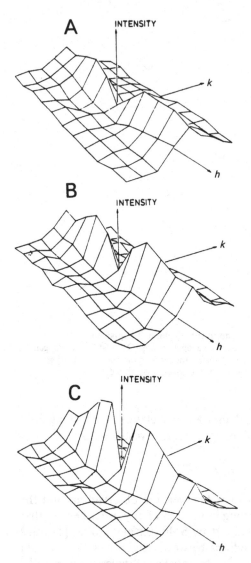

Figure 16. Elastic diffuse intensity in the $hk0$ plane [44] around the super-lattice point (0, 4, 1.5). The range of h is ± 0.5, while k ranges from 3.6 to 4.4. Three temperatures are: A, 273 K; B, 260 K and C, 250 K.

temperature [44, 45, 46]. As discussed before this gives information about local arrangements. If scattering occurs along a direction in reciprocal space then this is an indication that ordering fluctuations occur along a parallel direction in the crystal lattice; if the scattering has maxima then their location can, at least qualitatively, be used to estimate correlation lengths in the ordering process. The scattering in the $hk0$ plane will give information about arrangements in the *ab* projection, and this is shown in figure 16. Along the $h00$ direction there is strong scattering, while for $h = 0$ and especially along $0k0$ the diffuse intensity is very low. This indicates that no fluctuations occurred above the transition temperature along the polarization direction, but that transverse fluctuations of finite wavelength occurred along rows of the lattice. The scattering along $h00$ was found to peak at some distance from the origin of the reciprocal lattice [44], and this allowed the estimation of the full width of positive correlations in this direction to extend over four columns.

An explanation can be found in the conservation laws imposed by the ice rules. These require that polarization is constant from one row of the layer to the next [44], and consequently local fluctuations will be few until the transition occurs suddenly. Only along the rows can fluctuations occur with limited wavelengths, as described above. Similar considerations could now be used, with care, on ice-Ih. Again conservation rules are imposed [47]. Along c, for example, the bond polarization must be constant from layer to layer, and again this will tend to prevent local fluctuation, i.e. local higher order arrangement. One clearly has to be careful though, to extrapolate from copper formate tetrahydrate to ice Ih, both because the dimensions of the network are different, and because the presence of copper and formate ions are certainly going to change the bonding nature of hydrogen, and thus the mobility.

From the limited resemblance it is possible, though, to predict that a possible transition will occur at a lower temperature in ice-Ih than in copper formate tetrahydrate, and for several directions there can probably be no fluctuations prior to a possible phase transition. This is indeed in agreement with the observations quoted, that have found no or only slight indications of any ordering beyond what is prescribed by the ice rules.

4. Structural Implications of Spectroscopic Results

Ice-Ih, like all other solid and liquid phases of water, is understood to a large extent in terms of its hydrogen-bond properties. Spectroscopy has proven to be an extremely powerful tool in studies of hydrogen-bonded systems. The wealth of spectroscopic information on the ice phases well reflects its importance, and a full account of IR, Raman and neutron spectroscopy is far beyond the scope of this review. It does however seem to be quite useful to extract all structural information from the available spectroscopic data. For details of the dynamics involved the reader is referred to the literature (quoted partially in the following discussion) and we mention some recent reviews, namely on the full optical spectrum [48], the O–H stretching spectrum [49, 50] and on the neutron spectroscopy [51], all dealing at least in part with ice-Ih.

The mass difference between O and H(D) and the presence of strong hydrogen bonding helps considerably in the separation of translational, librational, bending and stretching parts of the spectrum. Furthermore, the mass difference between H and D on isotopic substitution allows a straightforward assignment of these well-separated broad bands. However, the detailed analysis and the definite assignment inside especially the librational, bending and stretching bands is still a matter of discussion. This is caused mainly by the fact that a crystal of ice has, due to its disorder, no strict local symmetry, which means that in principle 10^{24}–10^{25} modes are spectroscopically active in a given sample, and the usual normal coordinate calculations cannot be performed for such a system. One way to tackle the problem is to generate a reasonably big model of disordered ice and to sample the modes over all atomic arrangements occurring [52, 53]. Sometimes a more intuitive approach has been used to compare experimental spectra of ordered ices with the spectrum of disordered ice [54], which show some intriguing similarities; the spectroscopic resemblance of the stretch band has been confirmed in calculations on hypothetically ordered and disorded ice-Ih [53].

There are several ways of relating spectroscopy and structure. Depending on the electronic configuration, dipole moments and polarizabilities change compared with those of the isolated water molecule and tell us about the electronic rearrangements of water molecules in ice, which in turn provides information on structure and nuclear motion. Unfortunately a quantitative measurement of all IR and Raman intensities is still missing and only partial answers have been obtained [55]. The calculation of the atomic arrangements from the spectroscopically established force constants is another, still very rigorous approach, and some results will be presented here. Empirical structure frequency relationships can be established on the basis of well understood systems for which spectroscopic and crystallographic data are available (hydrates, ordered ices) and comparisons made with ice-Ih. Finally, the atomic thermal motion can be used as commutator between spectroscopy and diffraction data and some valuable structural information can be gained in this way. In the following sections we will discuss separately the different parts of the ice spectrum, starting with the high-energy side.

4.1. The Vibrational Band

It is well established that in ice-Ih the hydrogen atoms are not ordered over long distances. This implies that the vibrational motions of the hydrogen atoms vary according to the local arrangement. In addition to these disorder-induced effects, inter- and intramolecular coupling, interaction with external modes and anharmonic effects occur and influence the vibrational spectrum of ice-Ih. The position and shape of the stretch band is determined by the geometrical features of the hydrogen bond and the interaction between different internal as well as external modes. Since we are interested only in the geometric part, we have to try to separate these two effects. Following largely the interpretation of Bergren and Rice [52], the width and gross distribution of intensity in the stretch band is produced by strong coupling of the OH oscillators on different molecules, while the frequency of the stretch motion is accounted for by a variation in the harmonic force constants as a function of the hydrogen-bond-strength parameter η and by the anharmonic force constants of the free molecule, which are assumed to remain unchanged on hydrogen bonding.

The expressions for η are [52]

$$H_2O: \eta = -0.8569 + [2.0344 - 1.3007\,(\nu_{OH} - 10)/3711.46]^{\frac{1}{2}} \qquad (4.1a)$$
$$D_2O: \eta = -1.1183 + [3.0675 - 1.8169\,(\nu_{OD} + 5)/2729.39]^{\frac{1}{2}} \qquad (4.1b)$$

where $\nu_{OH}(\nu_{OD})$ are the experimental frequencies of the uncoupled stretch in cm^{-1}. The extra shift of $-10\ cm^{-1}$ $(+5\ cm^{-1})$ was introduced to correct for, the imperfect isolation of the OH(OD) oscillator in $D_2O(H_2O)$. The system (especially heavy ice-Ih) is further complicated by Fermi resonance between the stretching mode and the overtone of the bending mode. The influence of the long-range intermolecular interactions evoked by Whalley [55, 56] seems to have only very little influence on the intensity distribution within the stretch band. It is clear from the preceding discussion that only the uncoupled OH(OD) stretch frequencies of isotopically diluted HDO in $D_2O(H_2O)$ ice provide a sound base for intercomparison of spectroscopy and structure. A tabulation of experimental uncoupled stretch frequencies is given in Sceats and Rice [49]. However, even the uncoupled stretch has to be considered with some care. Johari and Chew [57] have shown by studying the pressure and temperature dependence of the coupled and uncoupled OH and OD stretching vibrations in ice-Ih that not only intermolecular coupling, the intermolecular distance and anharmonic effects determine the frequency; the temperature variation in the hydrogen bond strength at constant volume indicates that hydrogen-bond bending or changes in the anharmonic force constants occur and affect the uncoupled stretch. Moreover, since increasing the pressure does not mean shortening the hydrogen-bonded distance due to an increase in the bond strength [58], we have to expect a different behaviour for the stretch frequency on changing temperature, pressure or composition.

4.2. Stretch–Bond-length Relations

With the aforementioned intricacies it is obviously simpler to rely on the direct crystallographic information for the hydrogen-bond geometry, were it not for the problem of disorder hampering a straightforward analysis as discussed in section 2. So it was mainly because of some doubts about the crystallographically established O–H distance in ice-Ih that stretch–bond-length relations were considered as a useful source of information, for example [24, 59].

4.2.1. $\nu_{OH}(\nu_{OD})-r_{O-H}$ *Relations*. A very rough estimate of the covalent bond lengths r_{O-H} can be obtained using Badger's rule [60], which provides an empirical relationship between the force constant for bond stretching and r_{O-H}. In the harmonic approximation with the force constant K (in dyn cm^{-1}) given by

$$K = 4\pi^2 \mu c^2 \nu_{OH}^2 \qquad (4.2)$$

with the reduced mass μ of the OH oscillator, the speed of light and the stretch frequency ν_{OH} we have

$$r_{O-H} = 0.335 + (1.86 \times 10^5/K)^{\frac{1}{3}} \text{ Å} \qquad (4.3)$$

which gives

$$\Delta\nu_{OH}/\Delta r_{O-H} \simeq -13000 \text{ cm}^{-1} \text{ Å}^{-1}. \qquad (4.4)$$

There have been at least two other attempts [24, 61] to correlate ν_{OH} and r_{O-H} by differentiating O–H potential functions and a further two [11, 18] on the basis of experimental spectroscopic and crystallographic data. La Placa *et al.* [59] started from Libby's [62] potential constants. Later Klug and Whalley [61] rectified their result, and by adopting the potential function given by Smith and Overend [63] they obtain

$$\Delta_{O-H} = [99 - (9801 + 0.001348 - \Delta^2\nu)^{\frac{1}{2}}]/674 \text{ Å}$$

with $\Delta_{O-H} = r_{O-H} - r^0{}_{O-H}$
$$\Delta^2\nu = \nu^2{}_{OH} - \nu^{2}{}^0{}_{OH} \qquad (4.5)$$

where $r^0{}_{O-H}$ is the mean distance in water vapour and $\nu^0{}_{OH}$ the corresponding stretch frequency. The mean distances in the vapour for an HDO molecule are [64]

$$r^0{}_{O-H} = 0.9732 \text{ Å} \qquad (4.6a)$$
$$r^0{}_{O-D} = 0.9692 \text{ Å} \qquad (4.6b)$$

Note that the calculated r_{O-H} (r_{O-D}) has to be compared with the crystallographic bond length corrected for librational motions.

The construction of r_{O-H}/ν_{OH} diagrams allows the establishment of empirical, normally linear relationships. Novak [65] quotes from over 20 entries (mainly hydrates and organic compounds)

$$\Delta\nu/\nu_{OH}^0 = 3.6 \Delta r \qquad (4.7)$$

which gives for the uncoupled ν_{OH}

$$\Delta\nu/\Delta r \simeq -13350 \text{ cm}^{-1} \text{ Å}^{-1}. \qquad (4.8)$$

La Placa *et al.* [59] found from a comparison of HDO and ice-IX data for the uncoupled ν_{OD}

$$\Delta\nu/\Delta r \simeq -22000 \text{ cm}^{-1} \text{ Å}^{-1}. \qquad (4.9)$$

The calculated mean covalent bond length r_{O-H} at 123 K ($\nu_{OH} = 3276$ cm^{-1}, $\nu_{OO} = 2420$ cm^{-1}) varies between 0.987 (equation (4.9)) and 1.006 Å (equation (4.4)). The corresponding r_{O-D} values are assumed to be smaller by 0.004 Å. From these numbers the equilibrium distances r^e_{O-H} (r^e_{O-D}) can be obtained by subtracting the anharmonic shift due to zero-point motion.

Altogether these findings on their own do not seem extremely helpful for settling the issue of r_{O-H} in ice-Ih. Indeed, these relations have been used as evidence for the old crystallographic findings of long r_{O-H} in ice-Ih, questioning the r_{O-H} found in ordered ice-IX [61] or vice versa [24]. The best crystallographic r_{O-H} obtained, with oxygen disorder included, is only compatible with the shortest spectroscopically deduced r_{O-H}. Interestingly enough this estimate stems mainly from the best studied ordered ice-like structure known [59]. If the assumptions on linearity of the ν_{OH}/r_{O-H} relationships do not hold, or if the validity of the interatomic potential in water vapour for the ice phases is questioned, a comparison with a structure as closely related to ice-Ih should at least minimize the resulting errors. It is in this spirit that we have greatest confidence in equation (4.9) for a comparison with ice-Ih. We conclude that the true r^0_{O-H} and r^e_{O-H} in ice-Ih is still only approximately known and that spectroscopic results are only of limited use in checking the experimental evidence from crystallography.

4.2.2. $\nu_{OH}(\nu_{OD})$–$r_{O...O}$ *Relations.* For a long time the $r_{O...O}$ distances in ice-Ih were considered to being well established by crystallographic means, placing the oxygen atoms at the nodes of the perfect tetrahedral ice lattice. However, doubts about this picture were raised as early as 1964 by Bertie and Whalley [66]. They explained the halfwidth of the uncoupled stretch band in terms of a spread of $r_{O...O}$ originating from static displacements of oxygen by virtue of the proton disorder. They argued that the corresponding disorder component could not be detected by crystallographic means due to the dominating thermal smearing. Although the crystallographically-established mean value of $r_{O...O}$ was not questioned, it can certainly not be taken for granted that the distance remains unchanged on disordering. Further indication of this point using $\nu_{OH}(\nu_{OD})$–$r_{O...O}$ relations is therefore important, since crystallography does not give a definite answer (see section 2). The same relations are used to establish the spread in $r_{O...O}$ from the FWHM of the spectroscopic stretch band.

It is a well known fact that the relationship between stretch frequencies and hydrogen-bonded oxygen separations is not linear. Asymptotically approaching the stretch frequency of the free water molecule at large separations, it changes slope more or less continuously when starting from the short symmetric hydrogen-bond distances [65]. Most of the existing relations are purely empirical, although the mathematical functions are generally chosen to meet the asymptotic behaviour.

A frequently applied empirical relation is due to Falk [67], based on deuterated hydrates, high-pressure ices and some temperature-dependent data on ice-Ih

$$r_{O...O} = [20.96 - \ln(\nu_{OD}^0 - \nu_{OD})]/5.539 \text{ Å} \tag{4.10}$$

which gives, for $r_{O...O} = 2.76$ Å,

$$\delta\nu_{OD}/\delta r_{O...O} = 1610 \text{ cm}^{-1} \text{ Å}^{-1}. \tag{4.11}$$

More recently Berglund et al. [68], based on 37 data pairs for deuterated solid hydrates, quote

$$r_{O...O} = [16.009 - \ln(\nu_{OD}^0 - \nu_{OD})]/3.73 \text{ Å} \tag{4.12}$$

which gives again, for $r_{O...O} = 2.76$ Å,

$$\delta\nu_{OD}/\delta r_{O...O} = 1131 \text{ cm}^{-1} \text{ Å}^{-1}. \tag{4.13}$$

A similar relation for the uncoupled ν_{OH} was established by Efimov [69]

$$r_{O...O} = [16.916 - \ln(\nu_{OH}^0 - \nu_{OH})]/3.925 \text{ Å} \tag{4.14}$$

with $\delta\nu_{OH}/\delta r_{O...O} = 1720 \text{ cm}^{-1} \text{ Å}^{-1}.$ (4.15)

Moreover, Berglund et al. [68] give a relation between stretch frequency and the hydrogen-bond distance $r_{D...O}$

$$r_{D...O} = [11.156 - \ln(\nu_{OD}^0 - \nu_{OD})]/3.02 \text{ Å} \tag{4.16}$$

which will better account for the influence of the hydrogen-bond bending on the stretch frequencies. The scatter in the $\nu_{OD}/r_{D...O}$ plot is not noticeably reduced compared with the $\nu_{OD}/r_{O...O}$ plot [68]. In both cases the scatter is clearly larger than expected from the experimental error. This means that the effect of geometric bond bending is not a very important reason for the scattering in the data. From equation (4.16) we calculate a hydrogen-bond distance of 1.80 Å in D_2O at 123 K (1.79 Å at 60 K), which is clearly longer than the old crystallographic values obtained with fully ordered oxygen atoms (1.75 Å), but in agreement with recent calculations when oxygen disorder is included [23] (see section 2).

In order to answer the question of the mean $r_{O...O}$ in ice-Ih, we have to make a choice between the existing relations since they do not give the same answer. Falk's formula (equation (4.10)) gives the lowest $r_{O...O}$ and we find 2.753 Å at 123 K from the uncoupled stretch frequency. This is closest to the crystallographic value of 2.751 Å obtained in D_2O ice without allowance for oxygen disorder (see section 2.3). We note, however, that Falk's formula is based partially on ice-Ih data and will therefore tend to reproduce the values entered. From equation (4.13) we calculate $r_{O...O}$ as 2.761 Å for D_2O and equation (4.15) gives a $r_{O...O}$ of 2.758 Å for H_2O (compared with 2.752 Å from old crystallographic data), all at 123 K. Bearing in mind that the error estimated from the scatter of data in those empirical relations is not better

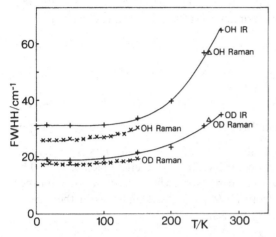

Figure 17. Temperature dependency of the FWHH of the uncoupled OH(OD) stretching vibration. Data from Sivakumar [70], (x); Logansen and Rozenberg [73], (+) and Johari and Chew [57], ($\triangle$). The estimated error for (x) is 4 cm^{-1} and less than 8% for (+). The lines are drawn as guides to the eye only.

than a few tenths of an ångström, a slight increase in $r_{O...O}$ due to the oxygen disorder, suggested by crystallography when oxygen disorder is included (see section 2.4), cannot be proved. We note, however, that, at least for H_2O, the difference between the crystallographic $r_{O...O}$ obtained with an ordered oxygen lattice and the spectroscopically estimated value is increasing with temperature. Such behaviour could be expected if the disorder in the oxygen positions is increasing.

The disorder of oxygen atoms and the resulting spread of $r_{O...O}$ has been invoked repeatedly, for example [57, 66, 70], in order to explain the residual width of the uncoupled stretch band shown in figure 17. The proton disorder creates local environments which potentially pull the oxygen out of its averaged (ordered) position with a resulting slight variation in $r_{O...O}$. The halfwidth of the $r_{O...O}$ distribution has been estimated several times using $\delta\nu_{OD}/\delta r_{O...O}$ relations given above. If we assume that hydrogen-bond bending is negligible and if we further assume that these relations describe equally well changes of the stretch frequencies with $r_{O...O}$ as a function of structural environment (from which they are established usually) as well as changes as a function of temperature, we obtain the increased full width at half height (FWHH) of $r_{O...O}$ with increasing temperature. The shape of the distribution has been discussed in some detail by Sivakumar et al. [70]. Noticing that anharmonic effects and bond bending could affect the stretch frequency and invalidate simple relationships, Johari and Chew [57] have tried to establish an independent estimate of this spread. Assuming that

$r_{O\ldots O}$ scales linearly with the lattice constants, they obtain from the measured $(\delta v/\delta T)_P$ and the experimental isothermal compressibility β

$$\left(\frac{\delta v}{\delta r_{O\ldots O}}\right)_P = \left(\frac{\delta v}{\delta p}\right)_P \frac{3}{\beta r_{O\ldots O}}\,\mathrm{cm^{-1}\,\mathring{A}^{-1}} \tag{4.17}$$

which gives at 255 K and 1 bar

$$(\delta v_{OH}/\delta r_{O\ldots O})_T = 755\,\mathrm{cm^{-1}\,\mathring{A}^{-1}} \tag{4.18a}$$

and $(\delta v_{OO}/\delta r_{O\ldots O})_T = 564\,\mathrm{cm^{-1}\,\mathring{A}^{-1}}$. $\tag{4.18b}$

With these relations they calculate the spread in $r_{O\ldots O}$ as 0.08 Å (H_2O) and 0.06 Å (D_2O) using the experimental FWHH of the uncoupled stretch. The choice of $(\delta v/\delta r_{O\ldots O})_T$ however seems somewhat arbitrary and another estimate could be obtained from $(\delta v/\delta r_{O\ldots O})_V$ calculated with the experimental dependencies at constant pressure

$$(\delta v_{OH}/\delta r_{O\ldots O})_P = 2958\,\mathrm{cm^{-1}\,\mathring{A}^{-1}} \tag{4.19a}$$

and $(\delta v_{OD}/\delta r_{O\ldots O})_P = 2042\,\mathrm{cm^{-1}\,\mathring{A}^{-1}}$ $\tag{4.19b}$

$$\text{or}\quad \left(\frac{\delta v_{OH}}{\delta r_{O\ldots O}}\right)_P - \left(\frac{\delta v_{OH}}{\delta r_{O\ldots O}}\right)_T = \left(\frac{\delta v_{OH}}{\delta r_{O\ldots O}}\right)_V = 2203\,\mathrm{cm^{-1}\,\mathring{A}^{-1}} \tag{4.20a}$$

$$\text{and}\quad \left(\frac{\delta v_{OD}}{\delta r_{O\ldots O}}\right)_P - \left(\frac{\delta v_{OD}}{\delta r_{O\ldots O}}\right)_T = \left(\frac{\delta v_{OD}}{\delta r_{O\ldots O}}\right)_V = 1478\,\mathrm{cm^{-1}\,\mathring{A}^{-1}}. \tag{4.20b}$$

The calculated spread in $r_{O\ldots O}$ is considerably reduced and using equation (4.20) it agrees better with the values from empirical $\delta v/\delta r_{O\ldots O}$ relations preserving however the slight difference between H_2O and D_2O. Which of these estimates is closer to the truth is still an open question. Empirical relations could be considerably in error for ice-Ih due to any reasons producing a scatter in the input data, while ice specific relations suffer from the uncertainty in the absolute scaling of $r_{O\ldots O}$ with the observed frequency shifts. Indeed, if we try to establish an experimental $v/r_{O\ldots O}$ relationship at 88 K, deduced from the coupled stretch [71], we obtain very high values for $(\delta v/\delta r_{O\ldots O})_V$ resulting in a negligible spread in $r_{O\ldots O}$, much smaller than expected from our crystallographic estimate. The agreement between the deduced $(\delta v/\delta r_{O\ldots O})_T$ and the empirical $\delta v/\delta r_{O\ldots O}$ relations is better and using the latter we calculate a spread in $r_{O\ldots O}$ which is comparable with the crude crystallographic value of a few hundredths of an ångström obtained at 123 K in D_2O (see section 2).

The FWHH of the uncoupled stretch does not change significantly below 100 K [70], which indicates that in this temperature range the spread of hydrogen bond strengths is a frozen-in property. Although this spread is usually expressed in terms of variations in $r_{O\ldots O}$, we should bear in mind that part of it could also be due to hydrogen bond bending. The importance of bending for the shift of the stretch frequency was pointed out by Johari and Chew [57]. Likewise, the increase in the FWHH at higher temperature is due to an increased spread in $r_{O \ldots O}$, an increase in the spread of bond hydrogen bond angles, or a combination of both. Moreover, a variation in the covalent bond lengths will result in a frequency spread, although in consideration of the force constants involved this seems to be of minor importance [57]. Finally, Scherer and Snyder [72] found in their single-crystal Raman work, slightly more linear hydrogen bonds parallel to the crystallographic c-axis compared with those oblique to the c-axis, which means that one expects some additional spread due to the different bond bending concerned with the two symmetrically independent hydrogen atoms. We conclude that simple $\delta\nu/\delta r_{O \ldots O}$ relations probably yield an oversimplified picture of the geometric arrangements and should always be taken as expressions for varying hydrogen-bond strengths only.

4.3. *The Vibrational Amplitude and Anharmonic Shift*

It is very useful to calculate the vibrational amplitudes from the observed stretch frequencies since they permit a straightforward comparison to be made with the crystallographically established mean-square displacements. Furthermore, the spectroscopic anharmonic shift may be related to positional changes of atoms seen in diffraction experiments and could help in that way to establish experimental equilibrium bond distances.

The relation between stretch frequency $\nu_{OH(OD)}$ and vibrational mean-square displacement $\langle u^2 \rangle_{OH(OD)}$ for an isolated harmonic O–H(O–D) oscillator in the low-temperature limit is given by

$$\langle u^2 \rangle_{OH(OD)} = h/8\pi^2 \, c\mu_{OH(OD)} \, \nu_{OH(OD)} \tag{4.21}$$

with the reduced mass $\mu_{OH(OD)}$ given as

$$\mu_{OH(OD)} = m_{H(D)} \, m_O/(m_{H(D)} + m_O) \tag{4.22}$$

where m_O and $m_{H(D)}$ are the atomic masses. Crystallographically we observe the difference in the vibrational components of H(D) and O

$$\langle u^2 \rangle_H = \langle u^2 \rangle_{OH} - \langle u^2 \rangle_O. \tag{4.23}$$

These components are obtained from equation (4.21) with $\mu_{H(O)} = m_{H(D)}$ or $\mu_O = m_O$. Table 6 gives the vibrational mean-square displacement for selected temperatures together with experimental results obtained from the analysis of heat capacity data in the quasi-harmonic approximation [27]; spectroscopic and heat capacity data are in excellent agreement. For the intercomparison with crystallography we refer to section 2.

Valuable structural information is contained in the spectroscopic study of anharmonic effects. Clearly every detail of the structure could be derived from the inter- and intramolecular potentials, if they were well known. Studies of the fundamental and overtone stretch bands give valuable insight

Table 6. Vibrational† mean-square displacements in ice-Ih

	Spectroscopy‡			Heat capacity§		
T (K)	$\langle u_V^2 \rangle_{\mathrm{OH(D)}}$ (Å)	$\langle u_V^2 \rangle_{\mathrm{O}}$ (Å)	$\langle u_V^2 \rangle_{\mathrm{H(D)}}$ (Å)	$\langle u_V^2 \rangle_{\mathrm{OH(D)}}$ (Å)	$\langle u_V^2 \rangle_{\mathrm{O}}$ (Å)	$\langle u_V^2 \rangle_{\mathrm{H(D)}}$ (Å)
H_2O 0	0.00545	0.000323	0.00512			
100	0.00544	0.000322	0.00511	0.00542	0.00032	0.00510
200	0.00540	0.000320	0.00508			
273	0.00537	0.000318	0.00505			
D_2O 0	0.00391	0.000437	0.00347			
123	0.00389	0.000435	0.00346	0.00404	0.00044	0.00360
223	0.00387	0.000433	0.00344			
273	0.00386	0.000431	0.00342			

† Displacements along the O–H(D) bond direction.
‡ Based on uncoupled IR stretch frequencies measured by Iogansen and Rozenberg [73].
§ From Leadbetter [27] corrected for anisotropy.

into these potentials in ice-Ih. The intramolecular potential is well established for the isolated water molecule and used as such for describing ice-Ih with allowance for some changes due to hydrogen bonding. Sceats and Rice [26] have discussed this approach in some detail. Here we are only interested in establishing the link between spectroscopic observation and its immediate structural implication. Characterizing the influence of hydrogen bonding by the strength parameter η given in equation (4.1) the decrease in the harmonic stretch frequency is given by

$$\omega(\eta) = \omega_0(1 - \eta) \tag{4.24}$$

where ω_0 is the gas-phase stretch frequency, which is equivalent to setting the diagonal stretch force constant given by Smith and Overend [63] to

$$k_{11} = 8.4452 \, (1 - \eta)^2 \text{ mdyn Å}^{-1}. \tag{4.25}$$

The anharmonic potential in the stretch direction is given by [26]

$$v_{\mathrm{anh}} = \frac{1}{r_{\mathrm{e}}} k_{111} \, (\varDelta r_1^3 + \varDelta r_2^3) + \frac{1}{r_{\mathrm{e}}^2} k_{1111} \, (\varDelta r_1^4 + \varDelta r_2^4) \tag{4.26}$$

where the subscripts 1 and 2 refer to the two covalent O–H bonds. The resulting anharmonic shift $X_{\mathrm{OH(D)}}$ is defined as the difference in the overtone and fundamental frequencies.

$$X_{\mathrm{OH(D)}} = (v_{\mathrm{O2}}/2) - v_{\mathrm{O1}}. \tag{4.27}$$

Assuming that $X_{\mathrm{OH(D)}}$ is dominated by the stretch components and that k_{111}

Table 7. *Anharmonic bond-length corrections from spectroscopy*

T (K)	X_{OH}^{3rd} (cm^{-1})	X_{OD}^{3rd} (cm^{-1})	Δr_{O-H}^{anh} (Å)	Δr_{O-D}^{anh} (Å)
60	−249.1	−136.1	0.0133	0.0096
123	−247.5	−135.0	0.0132	0.0096
223	−243.7	−132.9	0.0130	0.0095

and k_{1111} remain unchanged on hydrogen bonding, $X_{OH(D)}$ is given using equation (4.26) as

$$X_{OH(D)} = \tfrac{3}{2} k_{1111} \frac{1}{h c\, r_e^2\, \alpha^2(\eta)} - \tfrac{15}{4} k_{111}^2 \frac{1}{(h c)^2\, r_e^2\, \alpha^3(\eta)\, \omega_c(\eta)} \qquad (4.28)$$

with the harmonic oscillator function

$$2\alpha = 8\pi^2 \mu \omega_0(\eta)\, c/h. \qquad (4.29)$$

By inserting equation (4.27) and with the anharmonic constants [63]

$$k_{111} = -9.1577 \text{ mdyn Å}^{-1} \qquad (4.30a)$$
$$k_{1111} = 13.8390 \text{ mdyn Å}^{-1} \qquad (4.30b)$$

the anharmonic shift for isolated O–H and O–D oscillators is obtained as [26]

$$X_{OH} = 95.08\,(1-\eta)^{-2} - 170.49\,(1-\eta)^{-4} \text{ cm}^{-1} \qquad (4.31a)$$
$$X_{OD} = 50.39\,(1-\eta)^{-2} - 90.33\,(1-\eta)^{-4} \text{ cm}^{-1}. \qquad (4.31b)$$

These equations provide a satisfactory fit to the observed anharmonic shifts for various hydrogen bond strengths [26], and agree with an empirical relationship established by Berglund et al. [74]. In order to correlate anharmonic shifts with changes in the covalent bond length we obtain for a Morse-type potential in the limit of small Δr [75]

$$\Delta r_{O-H(D)}^{anh} = X_{OH(D)}/(2\alpha)^{\frac{1}{2}} \qquad (4.32)$$

where $\quad \chi_{OH(D)} = X_{OH(D)}^{3rd}/2\omega_0(\eta) \qquad (4.33)$

is assumed to originate in mechanical anharmonicity. Clearly only the third-order component of the anharmonic shift $X_{OH(D)}^{3rd}$ should enter into the calculation, obtained by setting the first term in equation (4.31) equal to zero. The anharmonic shift X_{OH}^{3rd} calculated for a Morse-type potential is larger by a factor 15/16 compared with the potential given in equation (4.26) which is negligible for our purpose. The resulting bond-length changes calculated with equation (4.32) and given in table 7 amount to magnitudes observable in diffraction experiments. The crystallographically measured distances have to be corrected for this effect in order to yield the experimental equilibrium bond length in ice.

We conclude the discussion of the vibrational band by stating that so far, in contrast to the stretch part, not much structural information has been gained from the analysis of the bending motion, since the observed bending frequency is only slightly increased compared with the vapour value. The conclusion can be drawn that the bending motions in ice are not much influenced by hydrogen bonding. The contribution of bending motions to the mean-square displacement of the H(D) atoms has not been taken into account in the calculation of the static disorder components perpendicular to the O–H(D) bond [22].

4.4. The Librational Band

A considerable effort has been devoted to the analysis of the librational band in ice-Ih. It is generally agreed that strong coupling between rotational modes causes the broadness of the band. The situation is further complicated by the uncertainties about the optical activities of the different librational modes. The magnitude of the translational–rotational coupling is not known, although in the case of some ordered high-pressure ices it has been found [77] that this effect is of minor importance. Clearly the librational band is affected by hydrogen bonding and a blue shift, indicating the tightening following the transition from water, is well established. However, a quantitative analysis has not yet been attempted, mainly because a detailed assignment of all features in the broad band is still missing. A tentative description of the band has been given by Sceats and Rice [25]. Assuming that the librations about the two-fold molecular axis are near the centre of the density of states, which is plausible because the moments of inertia is between those corresponding to the two remaining normal modes, and furthermore assuming that the twist motions, which are IR and Raman inactive in hypothetical ordered ice, contribute to the band, Sceats and Rice [25] approximate the librational density of states in H_2O by the set of frequencies

$$\nu_L^1 = \langle \nu_L \rangle + 144 \text{ cm}^{-1} = 954 \text{ cm}^{-1}$$

$$\nu_L^2 = \langle \nu_L \rangle - 17 \text{ cm}^{-1} = 793 \text{ cm}^{-1}$$

$$\nu_L^3 = \langle \nu_L \rangle - 127 \text{ cm}^{-1} = 683 \text{ cm}^{-1}.$$

These frequencies cannot easily be assigned to the rock, twist and wag motions. $\langle \nu_L \rangle$ denotes the centre of the band, which Sceats and Rice [25] locate at 810 cm^{-1} for H_2O. Recently the librational band observed with incoherent inelastic neutron scattering [78] has been explained as a convolution of three Gaussians centred at 935, 774 and 629 cm^{-1}, which is in fair agreement with the deconvolution of Sceats and Rice [25].

Based on these data and using different assignments, we calculate the associated mean-square amplitudes, again in order to establish correlations

Table 8. *Moments of inertia of a water molecule in ice-Ih†*

	I_{xx}	I_{yy}	I_{zz}
H_2O	1.235	1.858	0.623
D_2O	2.467	3.587	1.119
HDO	1.960	2.703	0.742

† Calculated for ice-Ih with an estimated $r^0_{O-H} = 0.98$ Å and $\gamma^0_{HOH} = 106°$; I is in amu Å². For the calculation of mean-square displacements perpendicular to the O–H(D) bond direction the proper geometrical factors have to enter. Note that the ratio of the mean-square displacements is equal to the ratio of the mean-square distances of the atoms from the principal axes of the molecule.

Table 9. *Librational mean-square displacements† in ice-Ih*

T (K)	H_2O (Å²)		D_2O (Å²)	
	$\langle u_L^2 \rangle_H$	$\langle u_L^2 \rangle_{H, anh}$	$\langle u_L^2 \rangle_D$	$\langle u_L^2 \rangle_{D, anh}$
60	0.01041	0.01067	0.00720	0.00739
123	0.01042	0.01081	0.00722	0.00748
223	0.01052	0.01126	0.00754	0.00812

† Averaged displacements perpendicular to the O–H(D) band. The reduced mass used in the calculation is 2 amu (4 amu) for H_2O (D_2O). The anharmonic frequency shift as obtained from Scherer and Snyder [72] as estimated from Leadbetter [27] is given as well.

with crystallographic results. With the moments of inertia for an idealized water molecule in ice-Ih given in table 8 and the well known relations

$$\langle \theta_L^2 \rangle = \frac{h}{8\pi^2 I_L \nu_L^0} \coth\left(\frac{hc\nu_L^0}{2k_s T}\right)$$

$$\langle u_L^2 \rangle = \frac{h}{8\pi^2 \mu_L \nu_L^0} \coth\left(\frac{hc\nu_L^0}{2k_B T}\right)$$

with ν_L^0 the librational frequency at 0 K, the mean-square angular (θ) and linear (u) displacements are obtained. The components of the displacements perpendicular to the O–H bond, calculated with assignments in terms of rock, wag and twist motions [76], show a variation in root-mean-square displacement larger than expected from the cylindrical symmetry of the hydrogen bond [79]. This cylindrical symmetry is reasonably well established from diffraction studies on crystallographically ordered water molecules [59]. Note however that wag and twist motions cannot be separated by diffraction; only the average of these out-of-plane motions is observed. In table 9 we give the total average of in-plane and out-of-plane components of the librational motions in ice-Ih. We quote this total average, since, as a consequence of the disorder, the crystallographically observed mean-square displacements perpendicular to the O–H bond are composed of in-plane and out-of-plane amplitudes. The adopted mean frequencies are based on the mean librational frequencies 810 cm^{-1} (585 cm^{-1}) for H_2O [25] (D_2O), in good agreement with experimental data [72, 78]. The librational mean square displacements obtained are smaller than those obtained from crystallography [22]. The difference can be used to calculate the disorder components perpendicular to the bond direction.

4.5. The Translational Band

Much can be learned about the dynamics of water molecules in ice from a study of the translational band in ice-Ih. The density of states and dispersion curves tell us a good deal about the crystal structure and the intermolecular arrangements. In a similar way to the O–H stretch, the translational frequencies yield information about the hydrogen bond strength, with respect to stretching or bending, and the width of the frequency bands is related to the distribution of bond lengths expected in a disordered system. The empirical correlation of frequency and bond-length is however still rather approximate [80] and at present not very helpful for a derivation of hydrogen bond distances and their distribution with reasonable accuracy.

The detailed discussion of dynamical aspects is beyond the scope of this review and we refer the reader to the literature [79–91]. Different approaches have been used to model the disorder in a lattice dynamical treatment of the translational band. It is established that the translational oscillations are significantly perturbed by the orientational disorder, although the gross features of the band are reconciled with H_2O molecules treated as point masses, whereby disorder mainly leads to a breakdown of the spectroscopic selection rules [81, 82]. However, in particular, the interpretation of features above 240 cm^{-1} necessitates the inclusion of hydrogen disorder either directly by introducing additional (mechanical) force constants [84, 85] or by using effective charges [86, 87], meant to describe indirectly the influence of the disordered hydrogen atoms. The considerable anisotropy of the force constants parallel and oblique to the c-axis invoked in earlier work [88] clearly is not required and is hardly justified for physical reasons. It seems that the crystal structure as seen in the translational band of ice-Ih is well represented by the interaction of disordered molecules centred at the averaged position of the periodic lattice. It is interesting to note that the isotopic ratio of the translational frequencies is smaller than expected from the mass ratio [27, 82, 84]. The difference cannot be ascribed to anharmonicity, but seemingly is due to slightly different effective intermolecular force constants [82]. This should be reflected in comparatively higher translational amplitudes in H_2O com-

Table 10. *Translational mean-square displacements in ice-Ih*

T (K)	$\langle u_T^2 \rangle_{H_2O}$† (Å)	$\langle u_T^2 \rangle_{H_2O}$‡ (Å)	$\langle u_T^2 \rangle_{D_2O}$† (Å)	$\langle u_T^2 \rangle_{D_2O}$‡ (Å)
60	0.0111	0.0139	0.0106	0.0127
123	0.0212	0.0231	0.0207	0.0220
223	0.0379	0.0420	0.0372	—

† Obtained from analysis of specific heat data [27]. The Debye temperature calculated from the same analysis gives $\theta_{D,H_2O} = 218 \pm 1$ K and $\theta_{D,D_2O} = 209 \pm 3$ K.

‡ Obtained from single-crystal neutron diffraction [23] with the static disorder components of oxygen subtracted as discussed in section 2.4. The difference between the specific heat and crystallographic data reflects molecular disorder as discussed in section 2.4 and given in table 5.

pared with D_2O. From the existing force constant models one could in principle calculate these amplitudes. However, this has not been done, and we refer to the analysis of the specific heat which yields the moments of the translational frequency spectrum. These moments can in turn be used to calculate the translational mean-square amplitudes [92] which are given in table 10. They are compared with the crystallographically oberved mean-square translational displacements. We note that the difference between $\langle u_T^2 \rangle_{H_2O}$ and $\langle u_T^2 \rangle_{D_2O}$ is small and difficult to verify in a diffraction experiment because it is obscured by pronounced atomic disorder. Even when the disorder component confined to the oxygen atoms and calculated in section 2 is subtracted out, a discrepancy between specific heat and crystallographic data remains, which probably is due to disorder involving the molecule as a whole. Such a disorder component is expected since there is a large variety of nearest-neighbour configurations involving slight molecular rearrangements. As seen in table 5, the molecular disorder is slightly larger in H_2O, which is in agreement with spectroscopic observations of the halfwidth of the O–H(D) stretch band and the corresponding spread in intermolecular distance, as discussed in detail above.

The flexing motion of three water molecules is responsible for the lowest frequency mode in the spectrum [25]. It could be of some importance for the calculation of the total mean-square displacement of oxygen with respect to hydrogen. However, the mean atomic displacements of the flexing motion are not known in detail and can therefore not be included in the averaged displacement of hydrogen with respect to oxygen perpendicular to the O–H band.

We conclude this section by noting that valuable structural information is obtained from a consideration of spectroscopy and crystallography via thermal motions, which first helps with the interpretation of diffraction data, and secondly provides some very stringent tests on their internal consistency.

For a full discussion we refer to section 2.4. Spectroscopy on its own will give direct access to the three-dimensional atomic arrangement only under very special circumstances and frequency/geometry relationships are seemingly not able to provide structural features with the required accuracy. A critical synopsis of all available data on the geometry of the water molecule and the hydrogen bond is given in section 8.

5. Structural Implications of Nuclear Resonance Data

A variety of nuclear resonance techniques has been used to probe the electromagnetic fields in the neighbourhood of atoms possessing a magnetic moment. In the following only the structural implications will be discussed. NMR work dealing with diffusional relaxation phenomena linked to the defect structure will not be considered and the reader is referred to the literature, for example [93, 94, 95].

Hydrogen, deuterium and ^{17}O have nuclear magnetic spins different from zero and thus a magnetic moment which could interact with the magnetic dipole field of electromagnetic radiation. In the case of hydrogen the relaxation is by a dipolar mechanism and the nuclear magnetic resonance (NMR) frequency corresponds to the energy splitting of the nuclear states in the presence of an external magnetic field. The electric field around the nucleus causes the local field to differ from the applied field. The induced diamagnetism lowers the field while the interaction with unpaired electron spins increases the field locally. The local field varies in different electronic environments and the NMR frequencies are shifted accordingly. This 'chemical shift' could be used to establish structure frequency relations, since diamagnetic shielding and paramagnetic anti-shielding depend on the actual charge distribution, which itself is a function of the hydrogen bond strength in the case of ice-Ih. Moreover, the nuclear magnetic moment is coupled to the surrounding molecules by the interaction of moving electrons. This interaction reduces the magnetic susceptibility with time, and the 'spin–lattice relaxation time' gives useful information about the intra- and intermolecular environment of a proton in the ice lattice. The relaxation time can be measured by following the time-dependent decay of the NMR signal following a radio-frequency excitation pulse ('free induction decay') [96] or by the analysis of the NMR line-shape [97] remembering that the line-width is proportional to the lifetime of the excited state. Free induction decay as well as the line-width of the NMR signal are characterized by the second moment M_2. The second moment, in turn, is related to the intra- and intermolecular H–H distances r according to the van Vleck formula [98]:

$$M_2 = 3\gamma^4\hbar^2 L(L+1)(4N)^{-1} \sum_{i \neq j} (1 - 3\cos^2\theta_{ij})^2 (r_{ij})^{-6}. \qquad (5.1)$$

γ is the gyromagnetic ratio, L the proton spin $(= \frac{1}{2})$ and N the number of protons in the sample. θ_{ij} is the angle between a vector r and the applied magnetic field. Equation (5.1) holds for a rigid lattice only and corrections for vibrational and librational motions have to be applied as discussed by several authors [96, 99, 100]. In order to extract specific H–H distances from the measured second moment the intra- and intermolecular contributions have to be separated:

$$M_{2,\,\text{total}} = M_{2,\,\text{intra}} + M_{2,\,\text{inter}}. \qquad (5.2)$$

Since the contribution to the second moment has an r^{-6} dependence, the intramolecular contribution dominates. By subtracting the intermolecular contribution, calculated usually by assuming a perfect tetrahedral half-hydrogen lattice and weighted with the appropriate probabilities, an experimental intramolecular second moment is obtained, which is used to calculate the intramolecular H–H distance with equation (5.1).

The second moment of hydrogen in ice-Ih has been measured repeatedly on powder and single crystals. Kume [97] calculated a range of $r_{\text{O–D}}$ distances and D–O–D angles, which were consistent with the second moments measured on a sample of powdered ice. Peterson and Levy's [17] results were found to be outside the allowed range. However, details of the calculation were not given and the measured moment was much higher than those measured in later work by Barnaal and Lowe [96]. These authors found agreement between the half-hydrogen model used in Peterson's and Levy's work [17], but other models were considered to be equally well supported by the data. A value of $M_{2,\text{inter}} = 12.83\ (\pm 0.11) \times 10^{-8}\ \text{T}^2$ has been calculated by Rabideau et al. [101], and Whalley [24] calculated from the measured total second moment. $M_{2,\,\text{total}} = 32.4\,(\pm 1.1) \times 10^{-8}\ \text{T}^2$, employing equation (5.1) and correcting for vibrational and librational motion an intramolecular hydrogen distance of 1.581 ± 0.016 Å. This value is significantly smaller than the value obtained from crystallography when the simple half-hydrogen model with tetrahedral bond angles is assumed (see section 2.3) and this discrepancy has been confirmed by later NMR work. Baianu et al. [102] have measured $M_{2,\,\text{total}}$: $32.1\ (\pm 2.0) \times 10^{-8}\ \text{T}^2$ and they calculated $M_{2,\,\text{inter}} = 9.6$ $(\pm 0.5) \times 10^{-8}\ \text{T}^2$. The resulting $r_{\text{H}\ldots\text{H}}$ is 1.58 ± 0.02 Å. Although the correction for vibrational and librational motion causes some uncertainty, it was definitely smaller than the difference between the crystallographically deduced distance and the NMR value. Based on this observation and supported by a series of other arguments, Whalley [24] concluded that the water molecule in ice is not much different from a water molecule in ice-II [103] and ice-IX [59] with the mean values of $r_{\text{H}\ldots\text{H}} = 1.58$ Å (ice-II and ice-IX, NMR), $r_{\text{D}\ldots\text{D}} = 1.567$ Å, $r^0_{\text{O–D}} = 0.982$ Å and $\measuredangle$ D–O–D $= 105.4°$ (all ice-IX, crystallography). Recent high-resolution neutron diffraction work by Kuhs and Lehmann [22, 23] is in agreement with the intramolecular hydrogen distances deduced from NMR data. Moreover, the NMR estimates of $r_{\text{H}\ldots\text{H}}$ are used to estimate the intramolecular bond angle, as discussed in section 8.

Nuclear electric quadrupole resonance (NQR) is observed when the electric quadrupole moment Q of a deuterium or ^{17}O nucleus interacts with the electric component of the electromagnetic radiation. NQR probes the electronic structure of a molecule much more directly than NMR does in that it gives the magnitude and symmetry of the electric field gradient at the position of the nucleus. The electric field gradient is readily calculated by *ab-initio* quantum chemical calculations and thus much more easily accessible to theory than the chemical shift. The deuteron quadrupole coupling e^2qQ/h, or more precisely, the deuteron field gradient q, depends predominantly on the O–D bond length [104–108]; the bond angle [109] and environmental effects play only a minor role in determining q_D. Hence deuteron quadrupole resonance data can be used to deduce structural parameters like the O–D bondlength r_{O-D} or, equivalently, the hydrogen-bond distances $r_{D...O}$ and $r_{O...O}$. Several empirical relations have been established [110–112]. Probably the most adequate relation for determining the hydrogen-bond distances in ice-Ih is based on data from water molecules in solid hydrates [68] and reads

$$r_{D...O} = (521/(303 - e^2Q/h))^{\frac{1}{3}} \text{ Å} \tag{5.3}$$

with a linear dependence of e^2qQ/h (in kHz) on $r_{H...O}^3$. By inserting the experimental NQR value for e^2qQ/h of 213.4 ± 0.3 kHz [113], a hydrogen-bond distance $r_{D...O}$ of 1.80 Å is obtained for D_2O ice-Ih. NQR experiments on diluted HDO in H_2O give a value of 213.9 ± 0.3 kHz for e^2qQ/h which means that the change in the hydrogen-bond length is only marginal when all neighbouring deuterons are replaced by protons. Likewise, Berglund *et al.* [68] have given an expression to relate e^2qQ/h to the hydrogen-bonded intermolecular distance

$$r_{O...O} = [13.67 - \ln(271 - e^2qQ/h)]/3.48 \text{ Å} \tag{5.4}$$

with the deuteron quadrupole coupling in kHz. Use of the experimental data quoted above yields an $r_{O...O}$ distance in D_2O of 2.763 Å and of 2.766 Å for HDO in H_2O ice, the difference being hardly significant. Altogether, NQR confirms the spectroscopically observed trend (see section 4.2) of a lengthening of $r_{D...O}$ and $r_{O...O}$ over the crystallographically established values based on a single half-hydrogen model. Considering the intramolecular r_{O-D}, unequivocal numbers have not yet been deduced for NQR data. Lindgren and Tegenfeldt [108] have established a relation between the shift of r_{O-H} and the shift of the deuteron quadrupole coupling e^2qQ/h based on *ab-initio* molecular orbital calculations. With the deuteron quadrupole coupling of the isolated water molecule, $e^2qQ/h = 307.9$ kHz [114], the shift is $307.9 - 213.4 = 94.5$ kHz, which corresponds to a very significant shift of 0.023 Å. With $r_{O-D}^e = 0.9585$ for the isolated water molecule the O–D distance in ice-Ih is then calculated as $r_{O-D}^e = 0.982$ Å. This theoretical estimate is in good agreement with results from other *ab-initio* calculations (see section 7). Davidson and Morokuma [115] on the other hand quote $r^e_{O-D} = 0.99$ Å in trying to reconcile NQR data with the old crystallographic r^o_{O-D}:1.011 Å, using large anharmonic shifts $\Delta r = r^o_{O-D} - r^e_{O-D} = 0.02$ Å derived from differences in the crystallographic bond lengths of H_2O and D_2O ice. This procedure however is barely justified, since the differences in bond length are to some extent due to disorder effects and do not solely originate in zero-point averaging (see section 2.4). They state that in order to be in agreement with the experimental field gradient ($q = 0.3183$ cm) the vibrational corrections due to zero-point motion have to be considerably different from the gas-phase values. By employing the gas phase corrections they obtain a r^o_{O-D} value which is about 0.01 Å shorter. This is in reasonable agreement with crystallographic values when oxygen disorder is included [22, 23], and certainly within the combined uncertainties of the zero-point averaging and inadequacies of the structure models used in the crystallographic analysis.

The electric quadrupole moment of a ^{17}O atom is insensitive to small changes in the local environment but extremely sensitive to overlap exchange and charge transfer [116]. Hence the structural information gained from ^{17}O NQR is expected to be small. From a fit to NQR spectra obtained from naturally abundant ^{17}O in ice-Ih at 77 K the following structural parameters were obtained [117]: $r_{O-H} = 0.99 \pm 0.02$ Å and $\angle H-O-H = 110 \pm 20°$. The errors are indeed too large to establish the geometry of a water molecule in ice-Ih with useful accuracy.

Altogether, nuclear resonance data are found to be in agreement with spectroscopic observation as concerns the different aspects of the structure of ice-Ih. They have corroborated the doubts on the crystallographically established interatomic distances and angles as obtained from the simplified half-hydrogen model. At present, NMR provides the only reliable source of information on the intramolecular hydrogen distance in the disordered ice lattice. To deduce from this number the intramolecular angle with a precision comparable with that achieved in crystallographic work on ordered structures, however, requires smaller uncertainties in $r_{H...H}$ than are at present available. Similarly, in order to deduce better estimates of r_{O-D} a careful intercomparison of NQR data of similar compounds could be very informative.

6. Computational Chemical Studies of the Geometrical Parameters

The main efforts of theoretical chemists interested in systems containing water have been to develop models and potentials to be applied to liquid water. In this effort the ices, with their known crystal structures, have mainly served as a test ground for different potentials, while ice-Ih in itself has only been studied in a limited number of cases.

6.1. Calculations with Fixed Molecular Geometry

Because of the cooperativity that occurs in hydrogen bonding the potentials involved must one way or another incorporate many-body interactions. This can either be done by a sum of effective pair potentials based either on central forces acting between pairs of atoms [118–120], atom–atom forces with additional Coulombic interactions between charges not at the atoms [121–123], possibly including polarizability [124], or potentials based on ab-initio calculations of water multimers [125–129]. An extensive test was made on different potentials [130] where these were used to predict molecular orientation and unit cell size for ice-Ih, -II, -VIII and -IX. The variables in this calculation were the unit cell, the centre of mass of the water molecule and its orientation expressed with Euler angles, while the covalent bond length O–H and bond angles were fixed at the value of an isolated water molecule. In this way the number of variables were 30 for ice-Ih and -VIII, and 78 for ice-II and -IX. The best result in terms of hydrogen-bond topology and O–O–O angles was obtained for the often used ab-initio potential by Matsuoka, Clementi and Yoshimine [129] (the MCY potential), but the density of ice-Ih was underestimated by 19.6% corresponding to an O...O length of around 2.96 Å rather than the experimental value of about 2.78 Å. The ab-initio potential in question [129] is obtained from energy calculations of a wide range of dimer arrangements followed by a fit of a suitable analytical function to these values. Contributions from three or more body effects are therefore not included, nor are the effects of changes in the water-molecule geometry, and it was therefore inferred that many-body components in the potential would act in a first approximation to reduce all lengths in the system by a uniform factor of 1.07.

Molecular orbital calculations have also been used to explain the differences between ordered and disordered hexagonal ice [131], the influence of the lattice forces on the water molecule geometry [132] and the different contributions to the lattice energy [133]. These calculations must be based on some model of a three-dimensional lattice, and this can be done either by imposing the same molecular orbital environment on all water molecules in the calculation [131], or by simply studying a large number of $(H_2O)_n$ clusters [134]. In both cases the calculations can become very complex, leading to compromises in the choice of method [131], and the problem of long-range interactions remains difficult to solve. It is interesting, though, to observe the differences in energy between the various configurations that can occur in the disordered lattice of hexagonal ice. If a cyclic cluster is taken out of an ice lattice then each water molecule is involved in two hydrogen bonds, and the individual water molecule might be one of three types – both hydrogen donor and acceptor, twice donor or twice acceptor [127] – and the distribution of the three types is 4:1:1. The optimal cycle in terms of energy is the sequential cycle [135], where all water molecules act both as acceptors and donors. By introducing donor–donor molecules or acceptor–acceptor molecules the energy of the system might increase. In a study of non-additive energy contributions in six-membered rings [134], expressed as the difference between the total energy of the system and the sum of the dimer energies, it was found that the introduction of a pair of double-donor, double-acceptor molecules would decrease this energy by a factor as large as 2.26, thus reducing the stability of this configuration. Calculations on large clusters showed similar effects, and this might eventually allow the determination of possible preferred configurations in a hypothetical ordered ice lattice. The calculations [134] also showed the importance of long-range water–water interactions, as even for a cluster of 48 molecules only 67% of the ice binding energy could be accounted for.

6.2. *Variation of the Water Molecule Geometry*

The calculations quoted above were done for idealized ice lattices with dimensions chosen in agreement with the experimental values, and with the water molecule in an optimized monomer geometry. Except in the study of Santry [132], where the geometry of the water molecule was varied, providing for a change both in the H–O–H angle from 104.7 to 110.5° and an increase in the O–H bond length by 0.009 Å, the studies were therefore not aimed at explaining the geometrical changes upon hydrogen bonding. It is, however, also essential to attempt to explain these features, especially in view of the contradictions that have shown up in the experimental data [24]. It is obvious though, in view of the results quoted above, that calculations will be difficult, as they must include both variation of the molecular geometry and the many-body effects. One approach is therefore to study small clusters of water by optimizing their geometry, while another possibility is to use both two- and three-body water–water potentials as well as an O–H bond potential and to vary the ice lattice for some ordered model of ice. Both methods have been used recently [135, 136].

As a part of a general study on hydrogen-bond phenomena Newton [135] studied clusters of water molecules, and used these to predict the expected lengthening of O–H and contraction of O...O that occur in ice-I. The calculations were closed-shell *ab-initio* calculations. Using a split-valence basis set (4-31G), all O...O distances in the multimer were kept to the same value, and only the hydrogen atoms involved in hydrogen bonding were shifted. Calculations of sequential hexamers, which is one ingredient of ice-Ih, gave O...O within 0.1 Å of the experimental value in ice, and extrapolations from monomer, dimer and hexamer O–H bond-length values could then be used to estimate the O–H length in ice when the experimental O...O length was used, and when it was assumed that the increase in O–H was proportional to the decrease in O...O. The expected O–H bond length in ice was in this way determined to be 0.988 Å.

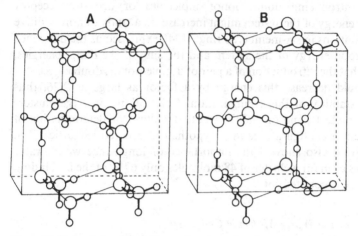

Figure 18. Ordered structures of ice used to determine the structural parameters [136]. (A) is ferroelectric and in crystallographic space group $Cmc2_1$, while (B) is anti-ferroelectric and in space group $Pna2_1$.

A complete variation of the ice lattice was carried out by Yoon, *et al.* [136] using two- and three-body potentials based on *ab-initio* calculations. Two ordered ice structures were used, and are shown in figure 18. In one case, the ferroelectric model, there are four molecules per unit cell while the other, anti-ferroelectric, has eight molecules per cell. The variables are the O...O and the O–H bond lengths, and there are three dimer and fourteen trimer configurations possible. Interaction energies, defined as the difference between the energy of the total configurations and the energy of its constituents, were then calculated for several sets of bond lengths, and potential functions were derived. A Hartree–Fock–Roothaan method was used, but for the geometries where the configurations for the MCY [129] potential are available, these were included. The trimer potentials were obtained as averages over the energy of the six trimer configurations that have a given water in the middle, and the twelve trimers, where the water is in an end position. For both dimers and trimers the variation in interaction energy was found to be approximately linear with the O–H stretch. Finally, a one-body potential [137] for the O–H stretch was included, and the lattice energy described as a sum of the outlined components was calculated for a range of values in O...O and O–H. The resulting binding energy expressed as the binding energy per molecule relative to the water monomer is shown in figure 19. For the two cases the O...O distances at equilibrium is between 2.79 (anti-ferroelectric) and 2.85 Å (ferroelectric), while the O–H is stretched to between 0.972 and 0.977 Å. Some difference can be observed in the binding energy, but it is believed that higher order terms will reduce this difference.

Both approaches described above have given results that account for the experimentally observed trends, and in general there is near quantitative

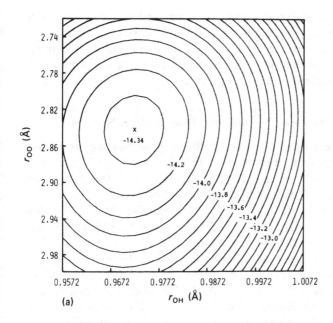

(a)

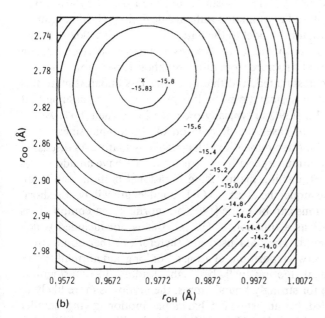

(b)

Figure 19. Binding as a function of O...O (r_{OO}) and O–H (R_{OH}) bond lengths in (a) ferroelectric ice and (b) in anti-ferroelectric ice [136]. Units are Å and kcal mol^{-1}.

agreement. Computations on systems as complicated as ice crystals still require large amounts of time, but there is obviously hope of obtaining very detailed theoretical understanding not only of the cooperative shrinking effects, but also of possible disorder effects at the oxygen position, and eventually on the preferred ordered arrangements.

7. Comparison with Hydrogen Bonding and Water in Other Systems

Many surveys exist of hydrogen bonding [138], and from these we can extract expected values for the quantities involved directly or indirectly in hydrogen bonding, namely the O–H covalent bond-length, the H...O hydrogen-bond contact, the similar O...O contact, the angle in the hydrogen bond O–H...O, and the H–O–H angle in the molecule. We can likewise obtain some idea from the large number of data on how likely a certain donor direction is with respect to the lone pairs of the oxygen atom.

As in the case of the general relationships derived from spectroscopic information, some caution must be exercised in the use of such relationships. Moreover, the results extracted are generally neither very precise nor accurate. Both experimental errors and errors of chemical interpretation are at play. To obtain good hydrogen positions neutron diffraction data must be used. This is in itself not a serious obstacle, but even then a typical good measurement does not lead to estimated standard deviations (ESD) obtained from the fit of the structural model to the data of better than 0.002 Å, when interatomic distances are in question. Moreover, intercomparison among data on the same crystal structure from different laboratories shows that the ESD tends to be underestimated, and that of course this underestimation is very much dependent on systematic error. It is probably safe to assume that the factor is at least 1.3 [139], so typically the precision is then 0.003 Å.

In a large majority of cases the description of the atomic smearing is assumed to be that of a three-dimensional harmonic oscillator. As discussed above, the true distribution, especially of hydrogen, does have an additional curvilinear component, coming mainly from librational motion, and when this is neglected in the model the observed O–H bond length will be too short. Approximate corrections can be made [140], but normally, in compilations of, for example, O–H as a function of H...O, this is not included which should be kept in mind. The data used are therefore neither equilibrium values nor distances between positions of atoms. For long and normally weak bonds, the librational motion will be high, leading to considerable foreshortening while for strongly bound atoms, the motion is more likely to be well approximated by an isotropic harmonic motion giving smaller systematic errors. In general the difference between the different positions is less than a few thousandths of an ångström.

A compilation of hydrogen-bond distance relationships is normally based on chemical compounds of a wide variety. A further spread in the observations

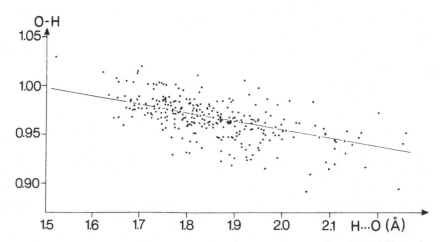

Figure 20. The relationship between O–H and H...O and regression line [145] with O–H = −0.109(7) H...O + 1.168. Stars are for non-molecular structures with the ESD of O–H less than 0.01 Å.

is therefore due to the variation in the chemical nature of the constituents. This can best be understood by observing that a hydrogen bond consists of a hydrogen donor and a hydrogen acceptor and can thus be seen as a Brøndsted acid–base interaction. In the hydrogen bond the hydrogen atom will be found at the weakest acid, and if the two groups are identical the bond will be short and symmetrical [141] or nearly so. Moreover, it would be natural to expect that not only the locations, but also the distances will be related to the acidity of the two groups. Although this has only been discussed in a very few structural analyses [142], a comparison of different hydrogen-bond systems clearly shows such a relationship, and many spectroscopic studies have been devoted to the study of relations between acidity/basicity and hydrogen bond strength [143]. In general the variation in relative pK will allow us to obtain relationships between O–H and H...O over a large range of H...O distances for a given fixed hydrogen-bond donor group O–H and variable acceptor group, but it might equally well lead to quite some scatter in the value of O–H for fixed H...O in the case where the O–H group is varied. This can happen when, for example, the lone pair of a water molecule is involved in metal coordination or strong hydrogen bonding. As it is impossible to estimate this scattering *a priori*, it is best therefore to deduce the error from the relationship itself.

The geometry of water molecules in crystalline hydrates both of organic (molecular) and inorganic compounds has been studied by Ferraris and collaborators [144, 145]. In all, 97 hydrates were analysed, and a series of trends occurs. For the O–H...O hydrogen bond the relationship between O–H and H...O is given in figure 20. For ice-Ih the H...O distance is within the range 1.76 to 1.80 Å depending on whether we use the mean oxygen

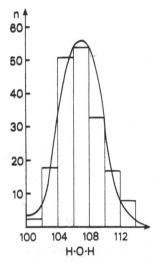

Figure 21. Histogram of H–O–H angles [145]. Mean value is 107.0° with a spread of 2.6° assuming a normal distribution.

position or the oxygen position suggested by the disorder model. In this range, therefore, the O–H bond length is expected to be 0.97 Å. The observed spread is quite large, ranging from values around 0.95 to 1.00 Å. The ESD from the linear regression is 0.01 Å, in agreement with this. A value for O–H well below 1.0 Å is therefore most likely, while the value for O–H in ice-Ih, assuming no oxygen disorder, is not entirely excluded.

In ice-Ih the bonding of oxygen is tetrahedral corresponding to an sp^3 hybridization of the atom. This would correspond to one of the ideal geometries of hydrogen bonding where two donors are placed along the two lone-pair directions of oxygen. Another likely bonding situation is one where the donor is directed along the bisector of the oxygen lone-pair (triangular coordination) and it has been proposed [146] that this would introduce more s character into the lone pair of the oxygen, and as a consequence lead to a larger H–O–H angle. This is indeed found [144, 145], and the mean values for triangular and tetrahedral coordination are 109.3° and 106.2°, respectively, both considerably larger than the angle of 104.5° observed in the gas phase [147]. Figure 21 shows a histogram representation of the experimental data, allowing for values of the angle of up to 112°. The water molecule in ice is therefore not atypical. It should also be noticed that there are of course other reasons for large H–O–H angles, such as the hydrogen bond forces and hydrogen–hydrogen repulsion which increases with the charge on the hydrogen atoms [144, 145] and is thus correlated with the strength of the hydrogen bond.

Altogether we can therefore conclude that the water molecule in ice-Ih is not atypical for a water bound in a solid. The range of O–H bond lengths would allow for a structure description where oxygen is not disordered, but clearly a shorter O–H, as obtained with an oxygen disorder model, would fit well with other O–H distances reported in the literature.

8. Summary

The structure of ice has very often been related to the structure of water and studies of ice are in many cases considered as preparatory for a better understanding of the liquid. The different phases of the water substance may be viewed as a basically tetrahedral network of hydrogen bonded water molecules, which is increasingly randomized when going from the crystalline state towards the vapour phase, passing through the amorphous and liquid states, for example [49, 50, 148, 149]. The phase changes are provoked by changes in the interaction between the water molecules and are manifested in changes of the atomic arrangements. One of the extremes of this series of phases, ice-Ih, was previously considered as fairly well understood [150, 151], although some dynamic aspects linked to defects of its structure and the mobility of its constituents were and are still under dispute. It is largely owing to Whalley, for example [24, 152] that doubts on the seemingly well-established structure of ice-Ih were kept alive and over the last few years our view of the structure of ice-Ih has changed considerably. The atomic arrangement in ice-Ih, established from a wealth of data discussed in the previous sections, is more accurate than previously claimed, but unfortunately less precise than one could wish. This is due to the fact that the atomic disorder is much more widespread than previously thought and crystallographic techniques have some difficulties in dealing with such systems. Nevertheless, by combining different experimental techniques, considerable insight into the structure of ice-Ih has been gained and in the following a synopsis of the individual results of the previous sections is given.

The oxygen disorder, well established from spectroscopy (see section 4.2) and crystallography (see section 2.4), has important consequences on the intra- and intermolecular geometry. The spatial site distribution has not been definitely established (see section 2.4), however, the displacement of the oxygen atoms into the bisectrix of the molecule is presently considered to be the most likely arrangement, as shown in figure 22. The oxygen disorder provokes an apparent increase in the covalent O–H(D) distances in the crystallographically established averaged structure (see section 2.3). This apparent increase was the main point of criticism in the past, but has now been completely resolved. All experimental techniques give reasonably close agreement for the O–H(D) bond length. The currently best estimate is $r^e_{OH(D)} = 0.973(5)$ Å, $r^o_{OH} = 0.987(5)$ Å and $r^o_{OD} = 0.983(5)$ Å, from a combination of crystallographic, NMR and spectroscopic data (the mean distances are given for a temperature of 123 K). The adopted H–H (D–D) distances

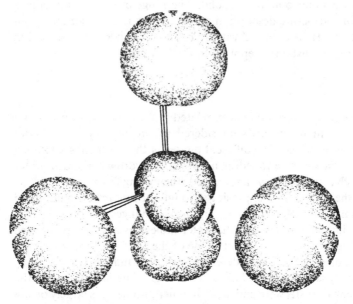

Figure 22. Schematic disorder pattern in one oxygen tetrahedron in ice-Ih. One of the possible configurations is shown with stick bonds.

are based on the NMR value of 1.58(1) Å, obtained from the second moments (see section 5). A direct measurement of the intramolecular angle is not available at present, but this angle may be deduced by combining the O–H(D) and H–H(D–D) distances quoted above, which yields 107(1)°. Thus, the intramolecular angle is clearly larger than the 'water angle' of the isolated molecule in the vibrational ground state (104.6°) and clearly smaller than the tetrahedral angle (109.47°) of the idealized ice lattice. Any difference in the equilibrium intramolecular geometry of H_2O and D_2O ice-Ih is smaller than the experimental errors.

These numbers fit quite reasonably to the values for water in other crystallographically well-established structures (see section 7). In this context a probably more interesting comparison is with ordered ice phases, since only these have a hydrogen-bond topology similar to ice-Ih; in ice-IX at 110 K [59] $r_{OD}^0 = 0.982(3)$ Å (corrected for thermal motion), $r_{D-D} = 1.567(3)$ Å and $\sphericalangle D–O–D = 105.4(6)°$ and in ice-VIII at 10 K [153] $r_{OD}^0 = 0.980(8)$ Å (corrected for thermal motion), $r_{D-D} = 1.56(1)$ Å and $\sphericalangle$ D–O–D $= 105.6(1.1)°$. The covalent bond-length is within the (large) limits of error identical in all thoroughly investigated ice phases, while the bond angle is marginally larger in ice-Ih when compared with the two high-pressure phases. All bond distances and angles in other, disordered high-pressure phases of ice, should not be compared directly with the ordered structures, since they probably exhibit a disorder of the oxygen atoms similar to that of ice-Ih or ice-VII [153].

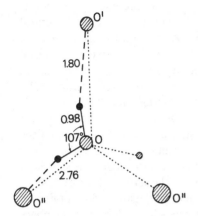

Figure 23. Schematic atomic arrangement in one oxygen tetrahedron in ice-Ih. The idealized hydrogen-bond directions are shown as dotted, the actual hydrogen bonds as broken lines. The numbers given correspond to the best estimate of distances (in Å) and angles (in degrees) obtained by combining all available data.

The hydrogen-bond geometry in ice-Ih as obtained when the oxygen disorder is allowed for, changes significantly. The hydrogen-bond distance $r_{H(D)...O}$, derived from crystallography, spectroscopy and NQR, is 1.80(1) Å, some 0.05 Å larger than previously thought. On the other hand, $r_{O...O}$ is only approximately 0.01 Å longer than the earlier value of 2.75 Å. The slightly longer $r_{O...O}$ in H_2O ice-Ih compared with D_2O, as obtained from NQR measurements, is hardly significant and needs further confirmation. For the temperature dependence of the hydrogen-bond geometry and especially $r_{O...O}$, an increase in length is expected from the blue shift of the spectroscopic stretch frequencies with increasing temperature. Using the $\nu_{OH(D)} - r_{O...O}$ relations discussed in section 4.2, one obtains an increase of 0.02–0.03 Å between 0 K and the melting point. The percentage increase of $r_{O...O}$ is significantly larger than the percentage increase of the lattice constants, thus indicating once more the existence of oxygen disorder in the ice-Ih lattice, which increases with increasing temperature. For the hydrogen-bond angles the picture is less clear at present. The hydrogen bond is obviously not exactly linear. There is, however, no experimental way of measuring these angles directly, so they have to be deduced from the interatomic distances with a corresponding accumulation of errors. Similar arguments hold for ∢O...O...O and we can only state that it is probably not far from tetrahedral. The best estimate of the intra- and intermolecular distances and angles are shown in figure 23.

From the above discussion it is obvious that this review would have been very different, had an ordered phase of hexagonal ice been at hand. It is probably too early to discuss all questions concerned with the possible disorder–order transformations of ice-Ih, since at present the matter is evolving quite rapidly. However, we shall briefly discuss some structural aspects involved in this phase transformation. We first notice that there is no evidence in available crystallographic data on pure ice-Ih that a structural change corresponding to a partial or full ordering occurs [19, 23]. This agrees with the fact that appreciable order in pure ice-Ih has never unequivocally been established by any other experimental technique, and the claim holds even in view of the fact that a ferroelectric ordering could take place with very few changes in interatomic distances and angles and thus very little change in the observed Bragg intensities. The majority of the theoretical approaches to the order–disorder phase transition in ice-Ih do indeed predict a ferroelectric ordering. Based on calculations of the static dielectric constant in the Kirkwood–Fröhlich [154, 155] approach, Minagawa [156] obtained a phase transition at 69 K. Hentschel [157] predicted a ferroelectric phase transition at 75 K based on the Onsager–Slater [158, 159] approach for the calculation of the dielectric constant. These predictions are in very good agreement with experimental observations on doped ice recently obtained by Suga and coworkers [160]. Neutron powder diffraction work on this ordered ambient pressure form of ice has been performed [161], but the data are not of the precision and accuracy necessary to help us any further in a better understanding of the disordered phase. It is interesting to note that the ordering of ice-Ih obviously cannot be achieved without introducing impurities or defects in the lattice [160]. This seems understandable from the finding discussed in section 3, namely that topological changes of the hydrogen positions necessarily involve rearrangements of the whole ice crystal. Jaccard [162] has proposed a theory of the order–disorder phase transition in ice-Ih, based on the kinetics of lattice defects. Work along these lines [163, 164] may give some better understanding of the thermodynamics underlying the phase transition. Ever since the days of Bjerrum [165] the nature of the lattice defects has been a field of active research of crucial importance for many properties of ice-Ih, not only in the context of the order–disorder phase transition but also for transport phenomena, ageing, etc. A thorough discussion of this subject is however beyond the scope of this review, although it appears that defects and lattice imperfections play a more important role in ice-Ih than in many other materials and merit a detailed discussion. Valuable information of this subject can still be found in the handbook on ice physics [166].

At present several studies remain to be done to clarify details of the structure of ice-Ih. First, we need to resolve the spatial arrangement of the disordered atoms in the ice lattice, and this can probably be done from further higher resolution diffraction studies at lower temperatures. Secondly, interatomic distances should be determined with higher accuracy, which can only be done by combining data from several experimental techniques. A comparison with the ordered (high-pressure) phases of ice will be helpful not only to confirm the observations made on ice-Ih, but also to supply a base for a better understanding of the molecular interactions in the ices, and eventually in the liquid phase. To proceed in this direction high-precision structural parameters in the high-pressure phases of ice remain to be determined. Obviously, as seen in the study of ice-Ih, the intercomparison with data from other techniques may be most helpful. Finally, we have still to understand the structural basis of a possible phase transition, and the interplay of the regular lattice with impurities and defects. Although these last quantities might not be measurable with diffraction techniques, which only give an averaged picture of the structure, an accurate and precise description of the ice-lattice is undoubtedly a sound basis for further progress.

Acknowledgements

We are indebted to Mrs R. Mason for preparing the manuscript and we thank Drs J. Axe and J. Schneider and Professors W. Descamps, G. Coulon, K. Morekuma and G. Ferraris for having supplied figures and information for this review. One of us (M.S.L.) thanks Dr and Mrs Jark for hospitality during part of the preparation of the manuscript.

References

1. J. Kepler, *Strena seu de Nive Sexangula*, Francofurti ad Moenum, apud Godefridum Tampach, 1611. (English translation: C. Hardie, *The Six-cornered Snowflake*, Oxford University Press, London, 1966).
2. V. F. Sears, *Thermal Neutron Scattering Lengths and Cross Sections for Condensed Matter Research*, Chalk River Report AECL–8490, 1984.
3. K. Lonsdale, *Proc. R. Soc. Ser.* A **247**, 424 (1958).
4. S. La Placa & B. Post, *Acta Crystallogr.* **13**, 503 (1960).
5. R. Brill & A. Tippe, *Acta Crystallogr.* **23**, 343 (1967).
6. H. Haltenorth, *Diploma – Thesis*, Technical University, München, 1968.
7. M. Blackman & N. D. Lisgarten, *Proc. R. Soc. Ser.* A **239**, 93 (1957).
8. G. P. Arnold, E. D. Finch, S. W. Rabideau & R. G. Wengel, *J. Chem. Phys.* **49**, 4365 (1968).
9. G. Dantl, *Phys. Kond. Mat.* **7**, 390 (1968).
10. J. Tatibuet, C. Mai, J. Perez & R. Vassoille, *J. Physique* **42**, 1473 (1981).
11. W. F. Kuhs, *Acta Crystallogr.* A **39**, 148 (1983).
12. D. W. J. Cruickshank, *Acta Crystallogr.* **9**, 757 (1956).
13. W. H. Barnes, *Proc. R. Soc. Ser.* A **125**, 670 (1929).
14. *International Tables for Crystallography*, Vol. A (ed. T. Hahn), International Union of Crystallography, D. Reidel, Dordrecht, 1983.
15. E. O. Wollan, W. L. Davidson & C. G. Shull, *Phys. Rev.* **75**, 1348 (1949).
16. L. Pauling, *J. Am. Chem. Soc.* **57**, 2680 (1935).
17. S. W. Peterson & H. A. Levy, *Acta Crystallogr.* **10**, 70 (1957).
18. R. Chidambaram, *Acta Crystallogr.* **14**, 467 (1961).
19. J. S. Chamberlain, F. H. Moore & N. H. Fletcher, in *Physics and Chemistry of*

Ice (ed. E. Whalley, S. J. Jones & L. W. Gold), Royal Society of Canada, Ottawa, 1973, p. 283.
20. W. F. Kuhs & M. S. Lehmann, *Nature* **294**, 432 (1981).
21. W. F. Kuhs & M. S. Lehmann, in *J. Phys. Chem.* **87**, 4312 (1983).
22. W. F. Kuhs & M. S. Lehmann, in *Colston Symposium on Water and Aqueous Solutions, April 1985, Bristol* (ed. G. W. Neilson).
23. W. F. Kuhs & M. S. Lehmann, *Acta Crystallogr. A* (1986).
24. E. Whalley, *Mol. Phys.* **28**, 1105 (1974).
25. M. G. Sceats & S. A. Rice, *J. Chem. Phys.* **72**, 3236 (1980).
26. M. G. Sceats & S. A. Rice, *J. Chem. Phys.* **71**, 973 (1979).
27. A. J. Leadbetter, *Proc. R. Soc. Ser. A* **287**, 403 (1965).
28. J. D. Bernal & R. H. Fowler, *J. Chem. Phys.* **1**, 515 (1933).
29. W. F. Giauque & J. W. Stout, *J. Am. Chem Soc.* **58** 1144 (1936).
30. P. Flubacher, A. J. Leadbetter & J. A. Morrison, *J. Chem. Phys.* **33**, 1751 (1960).
31. G. T. Hollins, *Proc. Phys. Soc.* **84**, 1001 (1964).
32. J. F. Nagle, *J. Math. Phys.* **7**, 1484 (1966).
33. J. F. Nagle, *J. Math. Phys.* **9**, 1007 (1968).
34. E. A. Di Marzio & F. H. Stillinger, *J. Chem. Phys.* **40**, 1577 (1964).
35. W. Marshall & S. W. Lovesey, *Theory of Neutron Scattering*, Clarendon Press, Oxford, 1971.
36. M. Descamps & G. Coulon, *Chem. Phys.* **25**, 117 (1977).
37. J. M. Rowe & S. Susman, *Phys. Rev. B* **29**, 4727 (1984).
38. P. G. Owston, *Acta Crystallogr.* **21**, 222 (1949).
39. J. Schneider, *Dissertation*, Technical University, München, 1975.
40. J. Schneider, & C. M. E. Zeyen, *J. Phys. C* **13**, 4121 (1980).
41. J. Schneider, & W. Just, *J. Appl. Crystallogr.* **8**, 128 (1975).
42. J. Villain & J. Schneider, in *Physics and Chemistry of Ice* (ed. E. Whalley, S. J. Jones & L. W. Gold), Royal Society of Canada, Ottawa, 1973, p. 285.
43. F. H. Stillinger & M. A. Cotter, *J. Chem. Phys.* **58**, 2532 (1973).
44. R. W. Youngblood & J. D. Axe, *Phys. Rev. B* **17**, 3639 (1978).
45. R. W. Youngblood & J. D. Axe, *Phys. Rev. B* **23**, 232 (1981).
46. R. W. Youngblood, *Ferroelectrics* **35**, 73 (1981).
47. J. F. Nagle, *J. Glaciol.* **21**, 73 (1978).
48. G. P. Johari, *Contemp. Phys.* **22**, 613 (1981).
49. M. G. Sceats & S. A. Rice, in *Water: A Comprehensive Treatise*, Vol. 7 (ed. F. Franks), Plenum Press, New York, 1982, chap. 2.
50. S. A. Rice, M. S. Bergren, A. C. Belch & G. Nielson, *J. Phys. Chem.* **87**, 4295 (1983).
51. H. Boutin & S. Yip, *Molecular Spectroscopy with Neutrons*, M.I.T. Press, Cambridge, Mass., 1968, chap. 9.
52. M. S. Bergren & S. A. Rice, *J. Chem. Phys.* **77** 583 (1982).
53. R. E. Shawyer & P. Dean, *J. Phys. C* **5** 1028 (1972).
54. E. Whalley, *Can. J. Chem.* **55**, 3429 (1977).
55. E. Whalley, *J. Glaciol.* **21**, 13 (1978).
56. E. Whalley & D. D. Klug, *J. Chem. Phys.* **71**, 1513 (1979).
57. G. P. Johari & H. A. M. Chew, *Phil. Mag. B* **49**, 647 (1984).

58. E. Whalley, in *The Hydrogen Bond* (ed. P. Schuster, G. Zundel & C. Sandorfy), North Holland, Amsterdam, 1976, chap. 29.

59. S. J. La Placa, W. Hamilton, B. Kamb & A. Prakash, *J. Chem. Phys.* **58**, 567 (1973).

60. R. M. Badger, *J. Chem. Phys.* **8**, 288 (1940).

61. D. D. Klug & E. Whalley, *J. Chem. Phys.* **81**, 1220 (1984).

62. W. F. Libby, *J. Chem. Phys.* **11**, 101 (1943).

63. D. F. Smith & J. Overend, *Spectrochim. Acta* A **28**, 471 (1972).

64. C. W. Kern & M. Karplus, in *Water: A Comprehensive Treatise*, Vol. 1 (ed. F. Franks), Plenum Press, New York, 1972, chap. 2.

65. A. Novak, *Structure and Bonding* **18**, 177 (1974).

66. J. E. Bertie & E. Whalley, *J. Chem. Phys.* **40**, 1637 (1964).

67. M. Falk, *Proc. Electrochemical Society Symposium on Chemistry and Physics of Aqueous Solutions*, Electrochemical Society, Toronto, 1975, p. 19.

68. B. Berglund, J. Lindgren & J. Tegenfeldt, *J. Mol. Struct.* **43**, 179 (1978).

69. Yu. Ya. Efimov, *Zh. Strukt. Khim.* **23**, 101 (1982), (English translation: *J. Struct. Chem.* **23**, 407 (1982)).

70. T. C. Sivakumar, S. A. Rice & M. G. Sceats, *J. Chem. Phys.* **69**, 3468 (1978).

71. G. E. Slark, A. K. Garg, W. F. Sherman & G. R. Wilkinson, *J. Mol. Struct.* **115**, 161 (1984).

72. J. R. Scherer & R. G. Snyder, *J. Chem. Phys.* **67**, 4794 (1977).

73. A. V. Iogansen & M. Sh. Rozenberg, *Opt. Spektrosk.* **44**, 87 (1978).

74. B. Berglund, J. Lindgren & J. Tegenfeldt, *J. Mol. Struct.* **43**, 169 (1978).

75. J. A. Ibers, *Acta Crystallogr.* **12**, 251 (1959).

76. M. Falk & O. Knop, in *Water: A Comprehensive Treatise*, Vol. 2 (ed. F. Franks), Plenum Press, New York, 1973, chap. 2.

77. G. Nielson & S. A. Rice, *J. Chem. Phys.* **80**, 4456 (1984).

78. J. F. D. Ramsay, H. J. Lauter & J. Tompkinson, Workshop on Water, April 1984, Institut Laue–Langevin, Grenoble, France. *J. Physique, Colloque* C 7, 73 (1984).

79. P. T. T. Wong & E. Whalley, *J. Chem. Phys.* **62**, 2418 (1975).

80. J. E. Bertie & D. A. Othen, *Can. J. Chem.* **50**, 3443 (1972).

81. E. Whalley & J. E. Bertie, *J. Chem. Phys.* **46**, 1264 (1967).

82. J. E. Bertie & E. Whalley, *J. Chem. Phys.* **46**, 1271 (1967).

83. J. E. Bertie, J. H. Labbé & E. Whalley, *J. Chem. Phys.* **50** 4501 (1969).

84. P. T. T. Wong, D. D. Klug & E. Whalley, in *Physics and Chemistry of Ice* (ed. E. Whalley, S. J. Jones & L. W. Gold), Royal Society of Canada, Ottawa, 1973, p. 87.

85. P. Bosi, R. Tubino & G. Zerbi, *J. Chem. Phys.* **59**, 4578 (1973).

86. D. D. Klug & E. Whalley, *J. Glaciol.* **21**, 55 (1978).

87. P. Faure, *J. Physique* **42**, 579 (1981).

88. B. Renker, in *Physics and Chemistry of Ice* (ed. E. Whalley, S. J. Jones & L. W. Gold), Royal Society of Canada, Ottawa, 1973, p. 82.

89. E. Whalley, in *Physics and Chemistry of Ice* (ed. E. Whalley, S. J. Jones & L. W. Gold), Royal Society of Canada, Ottawa, 1973, p. 73.

90. H. J. Prask, S. F. Trevino, J. D. Gault & K. W. Logan, *J. Chem. Phys.* **56**, 3217 (1972).

64   W. F. Kuhs and M. S. Lehmann

91. V. Mazzacurati, C. Pona, G. Signorelli, G. Briganti, M. A. Ricci, E. Mazzega, M. Nardone, A. de Santis & M. Sampoli, *Molec. Phys.* **44**, 1163 (1981).
92. T. H. K. Barron, A. J. Leadbetter, J. A. Morrison & L. S. Salter, *Acta Crystallogr.* **20**, 125 (1966).
93. J. A. Glasel, in *Water: A Comprehensive Treatise*, Vol. 1 (ed. F. Franks), Plenum Press, New York, 1972, chap. 6.
94. R. Blinc, H. Gränicher, G. Lahajnar & I. Zupančič, Z. *Phys.* B **22**, 211 (1975).
95. D. E. Barnaal & D. Slotfeldt-Ellingsen, *J. Phys. Chem.* **87**, 4321 (1983).
96. D. E. Barnaal & I. J. Lowe. *J. Chem. Phys.* **46**, 4800 (1967).
97. K. Kume, *J. Phys. Soc. Japan* **15**, 1493 (1960)
98. J. H. van Vleck, *Phys. Rev.* **74**, 1168 (1948).
99. B. Pedersen, *J. Chem. Phys.* **39**, 720 (1963).
100. B. Pedersen, *J. Chem. Phys.* **41**, 122 (1964).
101. S. W. Rabideau, E. D. Finch & A. B. Denison, *J. Chem. Phys.* **49**, 4660 (1968).
102. I. C. Baianu, N. Boden, D. Lightowlers & M. Mortimer, *Chem. Phys. Lett.* **54** 169 (1978).
103. B. Kamb, W. C. Hamilton, S. J. La Placa & A. Prakash, *J. Chem. Phys.* **55**, 1934 (1971).
104. T. Chiba, *J. Chem. Phys.* **41**, 1352 (1964).
105. M. Weissmann, *J. Chem. Phys.* **44**, 422 (1966).
106. M. Dixon, T. A. Claxton & J. A. S. Smith, *J. Chem. Soc., Faraday Trans. II* **68** 2158 (1972).
107. H. Huber, *J. Chem. Phys.* **83**, 4591 (1985).
108. J. Lindgren & J. Tegenfeldt, *J. Mol. Struct.* **20**, 335 (1974).
109. S. D. Gornostansky & C. W. Kern, *J. Chem. Phys.* **55**, 3252 (1971).
110. G. Soda & T. Chiba, *J. Phys. Soc. Japan* **26**, 249 (1969).
111. G. Soda & T. Chiba, *J. Chem. Phys.* **50**, 439 (1969).
112. M. J. Hunt & A. L. Mackay, *J. Magn. Reson.* **15**, 402 (1974).
113. D. T. Edmonds & A. L. Mackay, *J. Magn. Reson.* **20**, 515 (1975).
114. J. Veerhoeven, A. Dymanus & H. Bluyssen, *J. Chem. Phys.* **50** 3330 (1969).
115. E. R. Davidson & K. Morokuma, *Chem. Phys. Lett.* **111** 7 (1984).
116. D. T. Edmonds & A. Zussmann, *Phys. Lett.* **41A**, 167 (1972).
117. S. G. P. Brosnan & D. T. Edmonds, *J. Mol. Struct.* **58**, 23 (1980).
118. H. L. Lemberg & F. H. Stillinger, *J. Chem. Phys.* **62**, 1677 (1975).
119. A. Rahman, F. H. Stillinger & H. L. Lemberg. *J. Chem. Phys.* **63**, 5223 (1975).
120. F. H. Stillinger & A. J. Rahman, *J. Chem. Phys.* **68**, 666 (1978).
121. J. S. Rowlinson, *Trans. Faraday. Soc.* **47**, 120 (1951).
122. A. Ben-Naim & F. H. Stillinger, In *Structure and Transport Processes in Water and Aqueous Solutions* (ed. R. Horne) Wiley, New York, 1972.
123. F. H. Stillinger & A. Rahman, *J. Chem. Phys.* **60**, 1545 (1974).
124. P. Barnes, J. L. Finney, J. D. Nicholas & J. E. Quinn, *Nature* **282**, 459 (1979).
125. K. Morokuma & L. Pederson, *J. Chem. Phys.* **48**, 3275 (1968).
126. P. A. Kollman & L. C. Allen, *J. Chem. Phys.* **51**, 3286 (1969).
127. D. Hawkins, J. W. Moskowitz & F. H. Stillinger, *J. Chem. Phys.* **53**, 4544 (1970).
128. J. Del Bene & J. A. Pople, *J. Chem. Phys.* **52**, 4858 (1970).
129. D. Matsuoka, E. Clementi & M. Yoshimine, *J. Chem. Phys.* **64**, 1351 (1976).
130. M. D. Morse & S. A. Rice, *J. Chem. Phys.* **76**, 650 (1982).

131. K. T. No & M. S. Jhon, *J. Phys. Chem.* **87**, 226 (1983).
132. D. P. Santry, in *Physics and Chemistry of Ice* (ed. E. Whalley, S. J. Jones & L. W. Gold), Royal Society of Canada, Ottawa, 1973, p. 19.
133. E. S. Campbell & M. Mezei, *Mol. Phys.* **41**, 883 (1980).
134. D. Belford & E. S. Campbell, *J. Chem. Phys.* **80**, 3288 (1984).
135. M. D. Newton, *Acta Crystallogr.* B **39**, 104 (1983)
136. B. J. Yoon, K. Morokuma & E. R. Davidson, *J. Chem. Phys.* **83**, 1223 (1985).
137. G. D. Carney, L. A. Curtiss & S. R. Langhoff, *J. Mol. Spectrosc.* **61**, 371 (1976).
138. See for example: *The Hydrogen Bond*, Vols I, II, III (ed. P. Schuster, G. Zundell & C. Sandorfy), North Holland, Amsterdam, 1976.
139. J. J. Verbist, M. S. Lehmann, T. F. Koetzle & W. C. Hamilton, *Acta Crystallogr.* B **28**, 2006 (1972).
140. V. Schomaker & K. N. Trueblood. *Acta Crystallogr.* B **24**, 63 (1968).
141. J. C. Speakman, in *Structure and Bonding*, Vol. 12 (ed. J. D. Dunitz, P. Hemmerich, R. S. Nyholm, D. Reinen & R. J. P. Williams), Springer, Berlin, 1972, p. 141.
142. M. S. Lehmann & A. C. Nunes, *Acta Crystallogr.* B **36** 1621 (1980).
143. See for example: S. N. Vinogradov & R. H. Linnell, *Hydrogen Bonding*, Van Nostrand Reinhold, New York, 1971.
144. G. Ferraris & M. Franchini-Angela, *Acta Crystallogr*, B **28**, 3572 (1972).
145. G. Chiari & G. Ferraris, *Acta Crystallogr.* B **38**, 2331 (1982).
146. P. Coppens & T. M. Sabine, *Acta Crystallogr.* B **25** 2442 (1969).
147. K. Kushitsu, *Bull. Chem. Soc. Japan* **44**, 96 (1971).
148. J. A. Pople, *Proc. R. Soc. Ser.* A **205**, 163 (1951).
149. M. G. Sceats, M. Stavola & S. A. Rice, *J. Chem. Phys.* **70**, 3927 (1979).
150. D. Eisenberg & W. Kauzmann. *The Structure and Properties of Water*, Oxford University Press, London, 1969.
151. F. Franks, in *Water. A Comprehensive Treatise*, Vol. 1 (ed. F. Franks), Plenum Press, New York, 1972, chap. 4.
152. E. Whalley, *J. Glaciol.* **21**, 13 (1978).
153. W. F. Kuhs, J. L. Finney, C. Vettier & D. V. Bliss, *J. Chem. Phys.* **81**, 3612 (1984).
154. J. G. Kirkwood, *J. Chem. Phys.* **7**, 911 (1939).
155. H. Fröhlich, *Theory of Dielectrics*, Clarendon Press. Oxford, 1949.
156. I. Minagawa, *J. Phys. Soc. Japan.* **50**, 3669 (1981).
157. H. G. E. Hentschel, *Mol. Phys.* **38**, 401 (1979).
158. L. Onsager & M. Dupuis, in *Electrolytes* (ed. B. Pesce), Pergamon Press, New York, 1962. p. 27.
159. J. C. Slater, *J. Chem. Phys.* **9**, 16 (1941).
160. Y. Tajima, T. Matsuo & H. Suga, *J. Phys. Chem. Solids* **45** 1135 (1984).
161. A. J. Leadbetter, R. C. Ward, J. W. Clark, P. A. Tucker, T. Matsuo & H. Suga, *J. Chem. Phys.* **82**, 424 (1985).
162. C. Jaccard, *Phys. Kond. Mat.* **3**, 99 (1964).
163. M. Hubmann, *Z. Phys. B* **32**, 127 (1979).
164. M. Hubmann, *Z. Phys. B* **32**, 141 (1979).
165. N. Bjerrum, *K. Dan. Vidensk. Selsk. Mat.-Fys. Medd.* **27**, 1 (1951).
166. P. V. Hobbs, *Ice Physics*, Clarendon Press, Oxford, 1974.

Water Structure in Crystalline Solids: Ices to Proteins

HUGH SAVAGE

Center for Chemical Physics, National Bureau of Standards, Gaithersburg, MD 20899, USA and Laboratory of Molecular Biology, National Institute of Arthritis, Diabetes, and Digestive and Kidney Diseases, National Institutes of Health, Bethesda, MD 20892, USA

1. Introduction

The present picture we have of the three-dimensional structure of water in the majority of its existing states is far from satisfactory. Most of the structures in the solid phase – ordinary hexagonal ice-Ih, cubic ice-Ic and the high-pressure ices II to IX – are apparently well characterized, whereby the respective details are understood in the relatively simple terms of a tetrahedrally coordinated hydrogen-bonded lattice. The situation is very different for the liquid, aqueous solutions and amorphous ice. Within these phases, there is no underlying repeating lattice that forms the fundamental structural basis in crystals. The molecules can occupy many alternative positions relative to one another, provided they do not violate the inter-molecular forces: hydrogen-bonding, van der Waals contacts, dispersion, etc. Significant angular distortions from the expected tetrahedral coordination occur, but the extent and limitations of these deviations are unclear.

The main method used in analysing the details of molecular structure is diffraction (neutron, X-ray and electron). This has been invaluable for locating the individual relative atomic positions of the molecules in the crystalline state; however, for molecular assemblies which have no repeating units the method is more limited. Diffraction from liquids and amorphous substances yields only structural information about correlations between pairs of atoms, pair correlation function (*PCF*), which gives the probability of finding an atom within a certain distance of another. Some angular information is also obtainable from the most commonly observed distances (peaks in the *PCF*s). In addition to these difficulties, the molecules in liquids and solutions have mobilities that cannot be directly observed in diffraction experiments, since both time and space are sampled giving only an averaged picture of the structure.

To try to pin down what structural characteristics water must have, almost every available experimental technique has been utilized: for example, scattering methods (diffraction, inelastic), spectroscopy (IR, NMR, etc.) and calorimetry (scanning, heat capacity). These methods have been applied to both pure water and water–molecule complexes (e.g. crystals, aqueous solutions). Several reviews on this extensive area of research have been published [1–18] which critically assess and discuss the common properties and structural features of water obtained from the large number of experimental and theoretical results. To date, only a limited amount of structural information has been forthcoming, though strong suggestions are indicated by the nature of certain similarities between the results.

For these reasons, theoretical models have been widely used to fill the experimental void. Many different models have been proposed and developed, the numerous details of which are not discussed here but have been outlined and reviewed in several publications [3,4,13,15,19–25]. Generally, they are divided into two types: mixture and continuum models. The majority of the contemporary models are of the latter type, describing the water molecule using point charges, electropoles, dispersion, van der Waals forces, polarizability and sometimes other parameters such as fitting constants which may be used in conjunction with experimental or theoretical (quantum-mechanical) data. Although most models have been fairly successful in predicting many aspects of the macroscopic properties (thermodynamic functions, etc.), structural details at the molecular level remain unclear.

With most of the indirect methods remaining inaccessible in relation to probing the specific details of geometrical structure, we are forced to pay more attention to the main methods of structural analysis: diffraction. Crystal hydrates present excellent systems from which structural information about the geometrical consequences of water–water and water–c/molecule interactions can be obtained by X-ray and neutron diffraction. (c/molecule refers to the molecule(s) or ionic complexes that form(s) the actual crystal, be it an inorganic, organic or biomolecular crystal.) Hydrate structures range from small systems, such as amino acids and ionic complexes, containing a few water molecules to larger macromolecular systems, such as proteins, which may contain several hundred. In small hydrates the water molecules are usually fairly well ordered, often being integrated into the main hydrogen-bonding networks of the crystal. However, as the size of the c/molecule increases, water within the crystalline system is seen to become progressively more disordered. For example in protein crystals, which contain between 25 and 80 per cent solvent [26], most of the water is contained in solvent channels situated between the protein molecules. Except for close to the c/molecule surface, a large fraction of this water appears to be very disordered, apparently behaving diffusively very much like liquid water.

The success of modelling molecular assemblies in the condensed phases, as measured by the ability to reproduce experimental structural data, is at present somewhat limited. In earlier work on relatively simple systems, such as inert molecules and ionic crystals [27–33], the short-range repulsive contacts were seen to be the decisive factor in determining the final packing arrangements and individual positions of the atoms within the structures. To a first approximation, the repulsive contacts in these systems are treated as interactions between spherical atoms. Good agreement between predicted and experimental structures depends critically on the form and small variations of the repulsive functions. For example, ionic crystal structures are known to be very sensitive to the changes in the repulsive cores used – in some cases modelled as compressible atomic centres [32]. In water and aqueous phases, the strong attractive forces of hydrogen-bonding are generally believed to be the dominant factor in determining the structure. The repulsive forces are not usually considered to play such a decisive role. However, both the hydrogen-bonding and repulsive interactions vary quite rapidly over short distances and this suggests that the latter forces should not be overlooked. In this respect we may ask the following question: do we understand, at least at an intuitive level, which of the intermolecular interactions are involved in controlling the positioning of the atoms in an assembly of water molecules?

The basic structure of water in well characterized systems, such as crystal hydrates, has not as yet been adequately modelled. Different potential models have been used to simulate the water in several hydrates, but they appear to fall short of being able to predict *reasonable* positions and orientations for many of the water molecules that are very well localized [34–41]. Moreover, much emphasis has been placed on trying to derive the structural details of water from computer simulations, which in essence depends on how well the local and short-range interactions are represented by the potential used. The water molecule is *non-spherical* and the repulsive contacts are not well understood, mainly because of the complex nature of the interactions involved (electron cloud overlap, charge transfer, penetration effects). Hence, it would seem essential to ascertain the characteristics of its short-range structure and in this article we examine some of these aspects in known neutron crystal structures. Several regularities are found to arise, from which some interesting conclusions can be drawn that significantly increase our understanding of water structure(s) in general.

Section 2 outlines the main structural features (mainly hydrogen-bonding) that have been recognized for water in its various phases. The following section surveys the currently available crystal hydrates examined by neutron and X-ray diffraction. Subsequently, in section 4, details of the structural features of the short-range non-bonded interactions found in accurate neutron hydrate structures are described. The implications of short-range contacts within the various water phases are assessed in section 5. Then, in section 6, analysis of water structure in larger and more disordered hydrate crystals is addressed. Two main problems are seen to arise: (1) interpreting the regions of disordered solvent density and (2) extracting structural information from the sites assigned in the first stage. Inclusion of short-range structural information as restraints is found to be a very useful tool in the latter stage. The final section summarizes the main structural features and hopefully throws some light on how computer simulations can be better performed and interpreted using available short-range structural information.

2. Current Structural Information

The structural characteristics that have been well established [42–47] in both crystal hydrates and the ice polymorphs, mainly relate to the distances and angles, associated with the standard H-bond system:

$$X\text{–}H \cdots Y.$$

X and Y are electronegative atoms, of which one may be part of an adjacent c/molecule and the other a water oxygen, or they both may be water oxygens. Some information about close $X \cdots Y$ non-bonded next-nearest-neighbour contacts has been noted in the high-pressure ices. However, in most hydrate structures these contacts have not been analysed in detail. Short-range contacts, as we shall see later (section 4), are found to play an important role in the packing of waters within structures which contain significant distortions from tetrahedral H-bond coordination.

2.1. *Ice Polymorphs*

Eleven different forms of ice are known: ice-Ih (hexagonal), ice-Ic (cubic), and ices-II to X. Figure 1 shows the pressure–temperature phase diagram

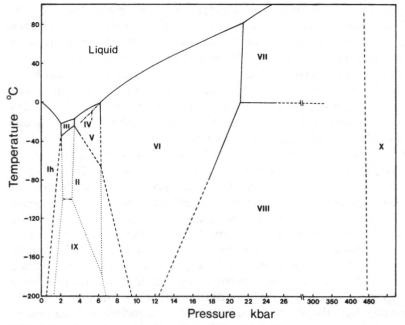

Figure 1. Pressure–temperature phase diagram of water. Full lines represent measured stable boundaries; chain lines represent measured metastable boundaries; broken lines represent presumed extrapolations of stable phases; dotted lines are metastable continuations of a phase into an adjacent region. Ice-Ic is not shown, but occurs in the lower part of the Ice-Ih region.

for water over the solid regions. Although both ices-Ih and -Ic exist at atmospheric pressure, the only natural form known to occur on earth is ordinary hexagonal ice-Ih: ice-Ic transforms irreversibly to ice-Ih. There is evidence that ice-Ic may occur at high altitudes in the atmosphere [48]. Ice II–ice-IX are formed at higher pressures of between 2 and 22 kbar, while ice-X exists at substantially higher pressures of 440 kbar and above [49]. The structure of ice X has not as yet been derived, but it is postulated to be of ionic character containing symmetric hydrogen bonds [50]. The hydrogens in ice-II and ice-VIII are fully ordered while in the remainder (apart from ice X) they are either partially or fully disordered. In several cases there is controversy as to whether the hydrogens are partially ordered or not, especially at lower temperatures: for example, ice VI [51–53].

Figures 2 and 5 show the basic crystalline structures for each of the ice polymorphs, except for ice-X. A summary of the ranges of the H-bond distances, $O \cdots O \cdots O$ angles, size of rings, etc., in these ice phases is listed in table 1.

Ice-Ih. Each oxygen is H-bonded to four nearest neighbours with $O \cdots O$ distances of 2.76 Å (varies by approximately 0.01 Å over 100 °C). The $O \cdots O \cdots O$ H-bond angles are all tetrahedral to within approximately 0.2°. The closest next-nearest neighbour $O \cdots O$ contacts are 4.5 Å and the structure is very open (figures 2(a) and 2(b). The covalent O-D bond lengths as deduced by neutron diffraction [54, 55], appear to be significantly longer (1.01 Å) than generally observed in the other ice structures (average = 0.98 Å).

Ice-Ic. The structure of this ice phase [56] is essentially similar to that of ice-Ih. Both ices contain layers of hexagonal rings in the chair configuration (figure 2(a)). The main difference lies in the connections between the layers of 'chair' rings: see the hexagonal rings with filled bonds in figures 2(a) and 2(b). In ice-Ih the hexagonal rings along the vertical *c*-axis are of boat form (layers are eclipsed with respect to vertical bonds: figure 2(b)), while those in ice-Ic are of chair form (staggered with respect to vertical bonds: figure 2(b)), and the overall structure is analogous to the cubic form found in diamond.

Ice-II. This is a fully ordered phase of ice [57, 58]. The structure (figure 2(c)) is composed of columns of almost flat hexagonal rings linked by H-bonds that are somewhat distorted from tetrahedral coordination. The $O \cdots O$ H-bonds range from 2.77 and 2.86 Å, the O-H $\cdots O$ angles from 166 to 178° and the $O \cdots O \cdots O$ H-bond angles are greatly distorted from tetrahedrality, 81–130°. One of the main structural features of this ice phase is a close intrusive next-nearest-neighbour contact of 3.24 Å which is *not* an H-bond.

Ice-III and ice-IX. These are basically the same structures except that the hydrogens are almost fully orientationally ordered in ice-IX (figure 3(a)) and fully disordered in ice-III, the higher-temperature form. The structures [59, 60] are composed mainly of five-membered rings (also some seven-membered).

72   *Hugh Savage*

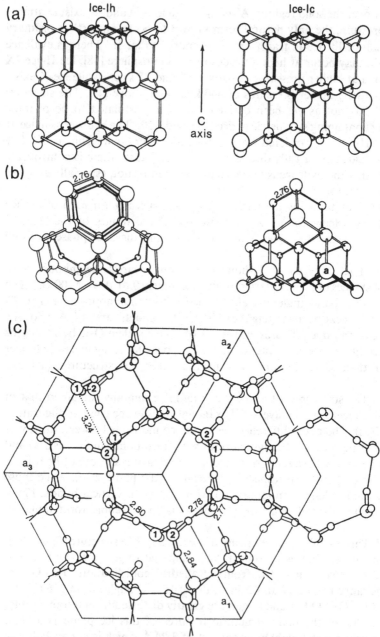

Figure 2. (a) Structure of ices-Ih (left) and -Ic (right) (oxygen positions) viewed perpendicular to the *c*-axis of the ice-Ih lattice. The filled bonds show the difference in the hexagonal ring conformations along the *c*-axis (of ice-Ih): boat form in ice-Ih and chair form in ice-Ic. (b) Structure of ices-Ih and -Ic viewed along the *c*-axis. The hexagonal rings are arranged in columns in ice-Ih, but are staggered in ice-Ic. H-bonded O···O distances are in ångströms. (c) Ordered structure of ice-II viewed along the hexagonal *c*-axis (the rhombohedral cell axes a_1, a_2 and a_3 are also shown). The H-bond O···O lengths and a close non-bonded O···O contact (dotted line) are given in ångströms [58].

Table 1

Ice polymorph	No of nearest neighbours and distances (Å) of H-bonds	No of next-nearest neighbours and distances < 3.6 Å	Hydrogen order	O–O–O angles (deg)	Density (g cm⁻³)	No of waters per unit cell	Ring sizes present				
							4	5	6	7	8
Ice-Ih	4, 2.76	—	Disordered	109	0.931	4			✓		
Ice-Ic	4, 2.75	—	Disordered	109.5	0.93	2			✓		
Ice-II	4, 2.77–2.84	9, 3.24–3.60	Ordered	81–128	1.18	12			✓		
Ice-III	4, 2.75–2.80	1, 3.45	Disordered	87–144	1.16	12		✓		✓	
Ice-IX(III)	4, 2.75–2.80	1, 3.45	Partial	87–144	1.16	12		✓		✓	
Ice-IV	4, 2.79–2.92	9, 3.14–3.29	Disordered[†]	88–128	1.27	16			✓	✓	✓
Ice-V	4, 2.76–2.80	7, 3.28–3.49	Disordered[†]	84–128	1.23	10	✓	✓	✓	✓	✓
Ice-VI	4, 2.80–2.82	17, 3.44–3.46	Disordered[†]	76–128	1.31	12	✓				✓
Ice-VII	4, 2.90	4, 2.90	Disordered	109.5	1.50	5			✓		
Ice-VIII	4, 2.88	5, 2.74–3.14	Ordered	109.5	1.50	5			✓		

[†] May be some partial ordering at low temperatures.

(a)

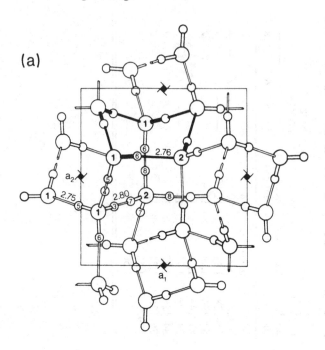

(b)

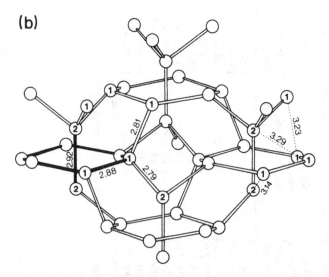

Figure 3. (a) Structure of ice-IX(-III), viewed in projection along the c-axis. The filled bonds show one of the five-membered H-bonded rings in the unit cell (axes a_1 and a_2 shown by box). Two networks of protons are present with occupancies of 0.96 (5, 6 and 7) and 0.04 (3, 4 and 8). O···O H-bond lengths are given in ångströms [60]. (b) Oxygen positions in the X-ray structure of ice-IV. The filled bonds represent a structural unit comprising an hexagonal ring with an H-bond passing through it. The H-bond lengths and close non-bonded O···O contacts (dotted lines) are given in ångströms [61].

(a)

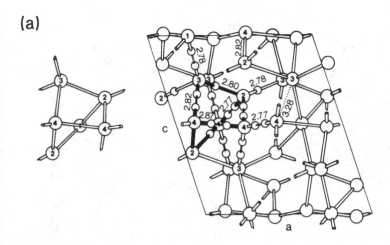

(b)

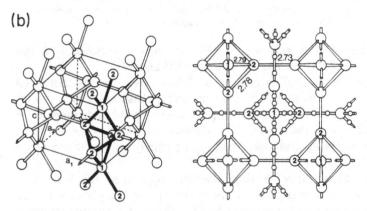

Figure 4. (a) Structure of ice-V viewed in projection along the *b*-axis (right). Some deuteron sites (partially occupied) are also included. The basic structural unit (left) comprises a four- and five-membered rings. The H-bond lengths and close non-bonded O···O contacts (dotted lines) are given in ångströms [65]. (b) Structure of ice-VI. One of the two interpenetrating lattices is shown on the left (filled bonds show a structural unit). On the right, the two lattices are shown in projection along the *c*-axis of the unit cell. The H-bond lengths are given in ångströms [64].

No hexagonal rings are present. The O···O H-bonds lie between 2.75 and 2.80 Å, while the O···O···O H-bond angles are more distorted at larger values than in ice-II, ranging from 91 to 144°.

Ice-IV. This is a metastable form that exists in the ice-V region of the ice *P–T* phase diagram (figure 1). The structure is interesting in that an H-bond is formed through the centre of an hexagonal ring (figure 3(b)). Only the X-ray structure is known [61] in which the O···O H-bonds are 2.79–2.92 Å and their O···O···O angles are between 88 and 128°. In this phase there are a

(a)   (b)

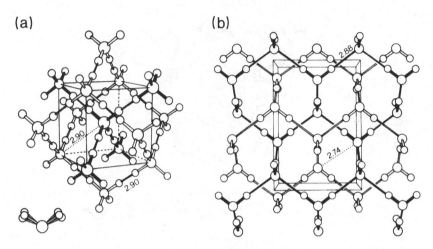

Figure 5. (a) Average positions in the structure of ice-VII. Below is shown a model for the disorder of the deuterons [67]. (b) Structure of ice-VIII [53]. The two interpenetrating lattices in each of these phases are shown by open and filled bonds. The H-bond lengths and short non-bonded O$\cdots$O contacts (dotted lines) in both structures are given in ångströms.

larger number of close non-bonded O$\cdots$O contacts (between 3.14 and 3.29 Å) than in all the other ices, except ices VII and VIII.

Ice-V. The structure [62] is composed of four-, five- and six-membered rings. Figure 4(a) shows the overall structure and the basic unit which is made up of 1 four- and 2 five-membered rings. The O$\cdots$O H-bonds range from 2.76 to 2.87 Å and the O$\cdots$O$\cdots$O H-bond angles from 84 to 128°. The protons in this structure were seen to be partially ordered in a neutron structural analysis [63].

Ice-VI. As the ice phases become more densely packed, this is the first phase to form two independently H-bonded interpenetrating lattices. Figure 4(b) shows the X-ray structure deduced by Kamb [64]. The neutron structure has recently been analysed by Kuhs *et al.* [53] and their results suggest that the protons are fully orientationally disordered in long range and are not partially ordered at low temperatures as indicated by earlier static permittivity experiments. [52] The O$\cdots$O H-bonds are 2.73–2.79 Å and the O$\cdots$O$\cdots$O angles are between 77 and 128°. The next-nearest-neighbour O$\cdots$O contacts are surprisingly long, with an average of 3.4 Å, compared with the short non-bonded contacts present in the above ices.

Ice VII and ice-VIII. These structures are the most densely packed forms of ice known that maintain the basic character of asymmetric H-bonding (the denser form of ice-X is thought to contain symmetric H-bonds with ionic character). They each comprise two interpenetrating (ice-Ic) lattices in which all the O$\cdots$O$\cdots$O H-bond angles are very close to the tetrahedral value. The neutron structure of ice-VIII has been examined by two groups [53,65] and was shown to be fully ordered (both oxygens and deuteriums). Each water molecule in the structure (figure 5(b)) is surrounded by eight nearest neighbours: four forming H-bonds and four making non-bonded contacts. The O···O H-bonds distances are 2.88 Å, longer than the two shortest O···O contacts of 2.74 Å. Three other short O···O contacts also occur, 3.03, 3.03 and 3.14 Å. In the X-ray structure of ice-VII [66] all of the eight nearest neighbours have the same distance of 2.90 Å. However, in recent neutron analyses of this phase [53,67] the results and interpretations indicate that both the oxygens and deuteriums (figure 5(a)) are probably disordered: approximately 0.1 Å for the oxygens and greater than 0.1 Å for the deuteriums. Ice-VII appears to be the disordered phase of fully ordered ice-VIII.

2.2. Hydrate Crystals

General surveys of the available small-molecule neutron structures [42,47] have indicated that (1) the X-H···Y angles of the H-bonds tend to be close to 180°, with the majority being greater than 150°, (2) the X-H···Y angles tend to decrease as the H···Y distances increase: see regression line (full) in figure 6. Non-covalent H···Y distances of less than the sum of van der Waals radii of H and Y are usually assumed to be H-bonds. When X and Y are oxygens, H···O distances of less than 2.4 Å are regarded as H-bonds, though very weak at distances longer than 2.0 Å.

Within the last decade the number of neutron hydrate structures analysed has increased rapidly, with over 250 such structures being available for a more detailed study. One of the more recent surveys by Chiari and Ferraris [46] included 97 highly refined hydrate systems in which the estimated standard deviations of the positional parameters were less than 0.02 Å. In this study, the majority of the hydrates were inorganic with most of the water molecules bound to ions. The well-known H-bond correlations were substantiated and a summary of the results is shown in figures 7 and 8. Geometrical parameters defined for donor and acceptor water molecules are illustrated in figures 7(a) and 8(a) respectively. Figures 7(b) to 7(h) show the distribution and average values of the standard geometries (covalent and H-bonds) for the water molecules. The covalent bond lengths and angles are fairly widely distributed with means of 0.96 Å and 107.0°. The O-H···Y bond angles tend to be strongly linear. For Y = O, the O···O distances range from 2.5 to about 3.1 Å (mean = 2.81 Å), while the H···O distances range from 1.5 to 2.3 Å (mean = 1.86 Å). The Y···O···Y H-bond angles range from 70 to 150° (mean = 107.6°); thus water molecules with angles outside this range may only form one reasonable donor H-bond.

The majority of donor atoms coordinated to the water molecules were observed to be cations and they apparently show a trend to be collinear with either the lone-pair directions or with the bisecting vector of the water molecule (Z-axis): see ω and ψ distributions in figures 8 (histograms shaded for C = hydrogen). There also appears to be a concentration of the coor-

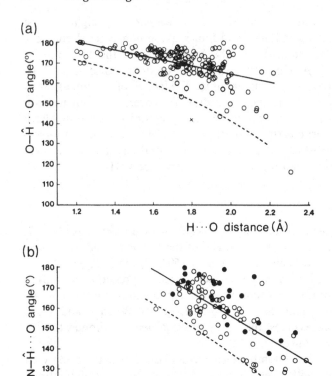

Figure 6. Angle X–H$\cdots$O as a function of H$\cdots$O distance for (a) O–H$\cdots$O bonds and (b) N–H$\cdots$O bonds in small structures that have been examined by neutron diffraction [47]. Only bonds with positional estimated standard deviations of less than 0.02 Å for the hydrogen were included. The full lines are regression lines for these plots; the broken lines represent the limits of H-bond bending.

dinated cations near the YZ plane (figure 8(d)), though for hydrogens (full circles) there is a wider scatter. The spread in the ω_1 and ω_2 angles is quite large (figures 8(b) and 8(e)), ranging from 0 to 100° (maximum at approximately 50°) and suggests that the directionality of the lone pairs to the donor atoms is not as significant as generally thought. This has been substantiated by quantum-mechanical calculations [68,69] from which there is evidence that the lone pairs are *not* well separated into two completely isolated regions.

Several workers have classified the coordination around water molecules into different groups with respect to the number and directionality of bonds to the lone-pair region. For example, Ferraris and Franchini-Angela [45]

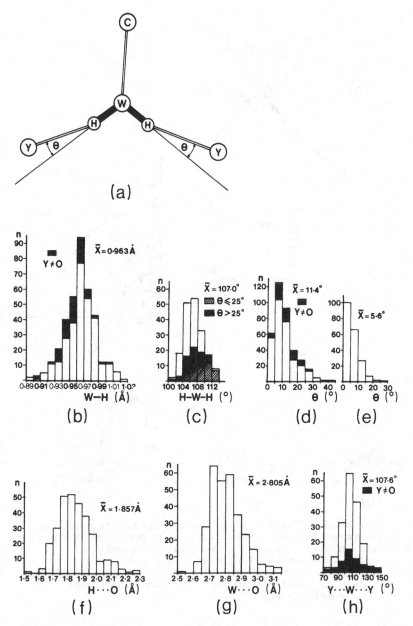

Figure 7. Characteristics of H-bonds donated by water molecules in crystal hydrates. (a) Parameters defined for the donor hydrogens: Y, the acceptor atoms; θ, the H-bond angle. Histograms (b) to (g) show the distributions of the various covalent (b and c) and H-bonding (d to h) geometries: (d) and (e) show the distribution of the number of hydrogen bonds *versus* the bending angle θ as observed, and after normalization to the unit solid angle respectively; (f) H$\cdots$O distances; (g) W$\cdots$O distances; (h) Y$\cdots$W$\cdots$Y angles [46].

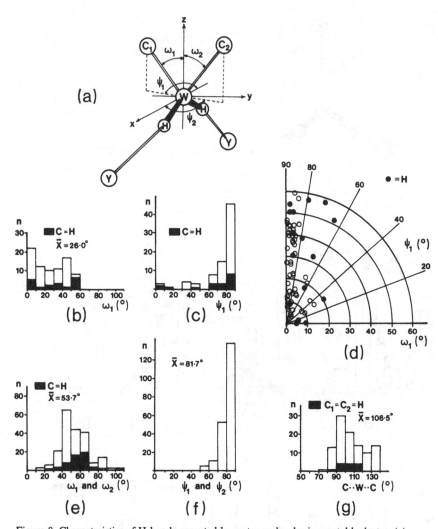

Figure 8. Characteristics of H-bonds accepted by water molecules in crystal hydrates. (a) Reference system around the acceptor region: C_1 and C_2 are the donor atoms (hydrogens or cations), ω_1 and ω_2 represent the angle between the C_1 and C_2 donors and the Z-axis (bisecting vector through the water molecule), ψ_1 and ψ_2 represent the angles of C_1 and C_2 around the Z-axis. Histograms (b) to (g) show various distributions of the quantities defined in (a): for the case of one coordinated cation (b–d) and for two coordinated cations (e–g). Shaded regions represent C = hydrogens [46].

defined five main types (in all classes the two hydrogens each form H-bonds): (1) one bond to the bisector of the lone pairs; (1′) one bond to one of the lone pairs; (2) two bonds to the lone pairs (tetrahedral arrangement); (3) three bonds to the lone pairs and (4) four bonds to the lone-pairs region. In the light of the quantum-mechanical calculations, there is some doubt whether these

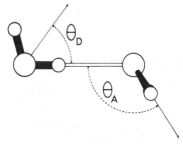

Figure 9. Definition of the orientational angles for a water dimer: acceptor angle θ_A and donor angle θ_D.

groupings are strictly meaningful since, with the lone pairs forming an elongated region of density, many different and effectively indistinguishable approach angles are possible. Other factors, such as short-range $O \cdots Y$ and $H \cdots Y$ repulsive contacts (discussed in sections 4 and 5), are also seen to play a significant role in determining the overall structural arrangements.

2.3. Gas Phase Dimer and Liquid PCF

Some information about bonding geometries and interatomic distances is also available from the gas and liquid phases. The experimental $O \cdots O$ H-bonding distance of the water dimer in the gas phase is 2.98 Å [70]. The acceptor and donor angles (defined in figure 9), are: $\theta_A = 123 \pm 10°$ and $\theta_D = 46 \pm 20°$ respectively. Several quantum-mechanical calculations have been performed on the dimer configuration [71–74] and the equilibrium $O \cdots O$ distances obtained vary over a fairly narrow range, between 2.92 and 3.02 Å. However, larger variations, in line with the experimental errors, are reported for the calculated θ_A and θ_D values.

Liquid scattering studies of water, using X-rays [75] and neutrons [76–78], provide information about the probabilities of finding an atom at a distance r from another at r_0: pair correlation function, $g(r)$. For water, the composite *PCF* is composed of three partials which correspond to the cross-pair correlations between the two atomic species present: $g_{OO}(r)$, $g_{OH}(r)$ and $g_{HH}(r)$. In the case of X-rays, the composite *PCF* is essentially that of the $g_{OO}(r)$ contribution (figure 10(a)), since hydrogens (compared with oxygen) are relatively weak scatterers of X-rays. For neutrons, all three partials are present, each with relatively high weights (figure 10(b)).

The positions of the peaks in the partial *PCFs* show the most probable (and spread of) distances between a particular pair of atomic species. The first peak in the X-ray $g_{OO}(r)$ is found at 2.85 Å (figure 10(a)), corresponding to H-bonding. The width of this peak ranges from a lower limit of about 2.5 Å to an upper limit that is difficult to identify because of overlap with the second peak, but is usually taken as about 3.3 Å (curve fitting). The second

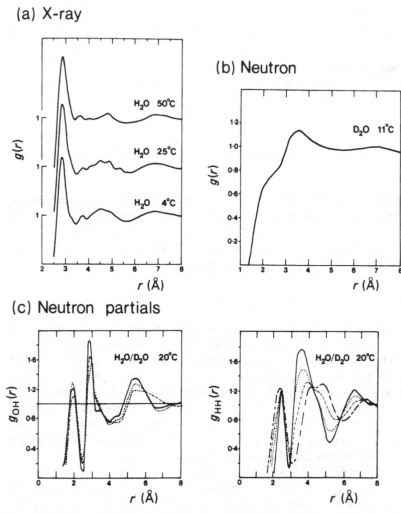

(a) X-ray

(b) Neutron

(c) Neutron partials

Figure 10. Pair correlation functions for liquid water: (a) composite X-ray *PCF* at three different temperatures [75]; (b) composite neutron *PCF*s [76]; (c) partial neutron *PCF*s: $g_{OH}(r)$ and $g_{HH}(r)$ functions calculated from different combinations of data sets [76].

$g_{OO}(r)$ peak, broader than the first, is centred around 4.5 Å and corresponds mainly to next-nearest-neighbour non-bonded contacts, similar to those in ices-Ih and -Ic. Thus, there appears to be some degree of tetrahedrality in the liquid structure. A third peak is also present at approximately 7.0 Å.

The neutron partials, $g_{OH}(r)$ and $g_{HH}(r)$ (figure 10(c)), appear to be somewhat difficult to separate out [76] and at present they are not generally as accurate as the X-ray $g_{OO}(r)$. In spite of this problem, analyses of neutron scattering data carried out by different groups [76–78], have shown that the first peaks of the $g_{OH}(r)$ and $g_{HH}(r)$ partials occur at relatively similar positions (differences in the order of 0.1 Å): $g_{OH}(r)$ at approximately 2.0 Å corresponding to H-bonding and $g_{HH}(r)$ at approximately 2.4 Å corresponding to H$\cdots$H repulsion. However, the agreement between the different analyses for the positions of the second peaks is not as good. Discrepancies in the order of 0.5 Å are observed.

2.4. Summary

In the ice polymorphs (excluding ice-X) the O$\cdots$O H-bond distances range from 2.73 to 2.92 Å and *all* the water molecules form four-coordinated H-bonded networks. In hydrate structures the O$\cdots$O H-bonds vary from 2.5 to approximately 3.2 Å which is a significantly wider range than in the ice polymorphs. The number of H-bonds formed per water molecule varies from two to six, with most being either three or four H-bond coordinated. Water molecules that participate in five or six H-bonds are usually assumed to be involved in bifurcated H-bonds (three-centred interactions). The wider spread of the hydrate O$\cdots$O H-bonds corresponds well with the range found in the first peak of the liquid X-ray PCF, 2.5 to approximately 3.2 Å – though it is not clear where the upper limit lies in the PCF since all bonded and non-bonded O$\cdots$O distances are present.

There is also a wider spread of the O$\cdots$O$\cdots$O H-bond angles in the hydrate structures than in ices: 70–150° in hydrates and 77–144° in ices. The larger range is probably allowed because of reduced repulsive strains between the oxygens of the longer H-bonds present in hydrates. Thus, the O–H$\cdots$O angles can become more distorted which in turn allows the O$\cdots$O$\cdots$O angles to increase or decrease. This is reflected in the observed ranges of O–H$\cdots$O angles (neutron analyses) in the ice and hydrate structures: 160–180° in ices and approximately 120–180° in hydrates. The H$\cdots$O distances of the hydrogen bonds also appear to have a wider distribution in hydrates than in ices: 1.5 to approximately 2.4 Å in hydrates (depending on the cut-off value used [42]) and 1.76 to 1.9 Å in ices.

Between the water dimer and the crystalline phases there is a considerable decrease in the O$\cdots$O H-bond distance, from 2.98 Å in the dimer to 2.76 Å in ice-Ih. A large proportion of this change can be attributed to attractive many-bodied forces such as polarization effects [79,80]. The experimental value of the dipole moment for a water molecule in the vapour phase is 1.85 D [81], yet in the condensed phases values of around 2.1–2.7 D have been estimated by various techniques [82,83]. An enhancement in the order of 40 per cent is apparent, due to polarization.

Essentially, the H-bonding geometries of water follow similar trends to those found in a majority of the H-bonding systems that have been surveyed [42–47], both in hydrates and non-hydrates. The structural features of the non-bonded O$\cdots$O contacts are, at present, not well characterized. In both the ice and hydrate structures, O $\cdots$ O contacts occur, down to approximately 3.1 Å. Such contacts probably also exist in the liquid structure, though here they cannot be observed directly. Most of the above analyses concentrate on geometries related to the attractive aspects of H-bonding, there being an inherent assumption that attraction controls distances and particularly angles. This assumption is questioned in sections 4 and 5 and the above hydrate and ice data are reassessed in the light of the role of repulsions.

3. Diffraction Studies of Hydrate Crystals

The two main experimental methods used in structural analysis are X-ray and neutron diffraction. General details on the theory and application of diffraction techniques are not discussed in this article, but can be found in references 84 to 87. The location of hydrogen atoms by means of X-ray diffraction is relatively difficult, since they scatter X-rays at substantially weaker levels (one electron) than other heavier atoms (e.g., C, N, O: six, seven and eight electrons respectively). Nevertheless, they can be located when accurate high-resolution X-ray data are available (better than 1 Å resolution), which is mainly the case for small structures. The level of accuracy obtainable is not as good as that for non-hydrogen positions (0.01 to 0.001 Å). When disorder is present, particularly in larger structures, the weaker electron density of the hydrogens is very often not observable.

The most reliable method for the detection of hydrogen positions is neutron diffraction. This is because hydrogen and its deuterium isotope are observed at comparable levels of magnitude to C, N and O atoms. However, there are some technical difficulties in undertaking neutron analyses. In contrast to X-ray sources, there are fewer experimental neutron facilities available, the running costs are higher and the obtainable flux levels are lower. Hence, only a fraction of the known hydrate crystals have been analysed by neutron diffraction (at present in the order of 250). Most of these structures are relatively small containing only a few water molecules.

There are also several problems arising in both diffraction methods with respect to the size of the c/molecular hydrate and to the attainable resolution of the data. This is particularly the case for medium and larger hydrate systems in which the solvent is frequently found to be disordered.

3.1. *Problems of Diffraction: Resolution and Size of Hydrate*

As noted in the introduction, hydrate crystals range from small structures containing relatively few water molecules (e.g., amino acids), to those which may contain several hundred. Although there is a large number of small hydrate crystals available, the relatively small number of water molecules present within these structures (often less than five), results in very few water–water contacts. Therefore, any analysis of the water–water interactions is somewhat restricted. In principle, the larger molecular crystal hydrates such as proteins and DNA offer very good systems in which the interactions between many water molecules can be studied, both in the bulk solvent regions and at the c/molecular interface. Unfortunately though, there are several problems which hinder us from deriving all the useful information. These are related to (1) the size of the hydrate system, (2) the resolution of the available diffraction data and (3) the disorder within the hydrate crystal of the c/molecules and especially the solvent (diffusive motions, degree of occupancies) which is often located in large channels. All three of these problems are interrelated and the following points can be made.

(a) In analyses of small and medium-sized structures, the resolution is usually sufficiently high (1.0 Å or better) to reveal individual peaks of electron or neutron scattering density for all the atoms present – both c/molecules and solvent. However, as the size of the system increases, the attainable resolution of the data is usually lower; between 1.5 and 2.5 Å for larger proteins (MW greater than 30 000). This is because of increased disorder of the c/molecule and also lattice disorder. At lower resolutions, the noise level in the electron/neutron density maps is often quite high and it becomes more difficult to differentiate, unambiguously, between noise and signal when assigning solvent positions. Nevertheless, for some smaller proteins, X-ray data have been obtained to atomic resolution and the background levels are seen to be significantly reduced.

(b) In larger systems one or more layers of hydration may be present and the first shell usually appears as fairly well-ordered peaks, with the water sites making many H-bond contacts to the molecule. In several cases, parts of the second and third shells may also be identified. In the layers further away from the surface of the c/molecule the solvent density becomes weaker and more diffuse, eventually merging with the featureless background continuum of the bulk solvent regions. Interpretation of these latter regions is usually almost impossible.

The observed density in a diffraction experiment represents a time- and space-averaged picture of the overall structure. The disordered solvent regions have to be modelled in some way. The most common method is to include partially occupied sites representing the amount of time spent there by solvent molecules, or alternatively, the probability of finding them there. This type of model is usually applied for cases of static disorder. Where dynamical disorder appears to be present, then a more complex model has to be used to account for the movements of the atoms/molecules. This is by no means a trivial problem even for a fairly rigid small molecule, since many variations of the atomic motions may be present.

The disentanglement and formulation of possible instantaneous water networks presents another major problem in deriving details of the structure from the solvent density (see section 6.2). This is mainly because there are very few reliable stereochemical restraints that can be applied to noncovalently bonded systems. The restraints for H-bonded systems appear to be more flexible compared with covalently bonded structures such as organic molecules.

3.2. *Groupings of Hydrates*

With respect to the size of the system and the resolution of available diffraction data, crystal hydrate complexes can be approximately divided into three main groups:

(a) Small hydrates: resolution less than 1.0 Å, containing less than 100 atoms.

(b) Medium hydrates: resolution $\simeq 1.0$ Å or better, containing 100–400 atoms.

(c) Large hydrates: resolution greater than 1.0 Å, containing more than 400 atoms.

These divisions are only presented as a guideline as some *overlap* exists between the groups. For example, several relatively small protein structures (crambin, BPTI, APP, see table 3) which fall within group (c), have been analysed to 1.0 Å resolution and below. With reference to both X-ray and neutron diffraction the following points may be made about each group.

Water molecules in small hydrates are usually present with unit occupancies in well-defined ordered positions. In medium-sized systems, disorder of both the c/molecule and the solvent begins to present problems. In large hydrates only the solvent that lies close to the c/molecular surface is fairly well ordered and interpretable.

3.3. *X-ray Diffraction*

Although a large amount of pertinent information about water structure is potentially available from all three classes of hydrates, very little has been derived from such X-ray studies that have thus far been attempted [88, 89]. This is partly due to difficulties in analysing solvents in medium and larger hydrates and partly due to the water hydrogen positions being somewhat inaccessible to X-rays. The hydrogen positions are required in order to assess the H-bonding networks within a structure. Thus, most X-ray analyses of water structure focus on the relative positions of the oxygens and other heavy atoms, from which only a limited amount of structural information is obtainable. Hence, for these reasons either the hydrates have to be examined by neutron diffraction, or an adequate water model may be employed to interpret the X-ray solvent electron density. In the light of what is known about water structure from smaller hydrates (sections 4 and 5) and also of the increased complexities in interpreting neutron solvent densities in larger hydrates (section 6.2), the latter method or a combination of both methods would appear to be valid approaches.

3.3.1. *Small Hydrates.* A large number of these structures have been solved with X-rays (probably more than 1000). In several of the ionic hydrates, theoretical hydrogen positions have been predicted using electrostatic and van der Waals models [90]. However, comparison with neutron diffraction data showed that, while agreement was fairly good for some structures (deviations of approximately 0.05 Å), in others the agreement was only moderate. These results indicated that the individual values used for the parameters (charges, etc.) of a given model, were probably structure specific. In addition to this, some of the basic models used were probably inadequate, for example, van der Waals interactions were often omitted.

3.3.2. *Medium Hydrates.* With atomic resolution data available, hydrates containing between 100–400 atoms in association with 10–50 water molecules (per c/molecule) have been studied. Very few of these hydrates have been analysed in detail with respect to the actual associated solvent structure (H-bonds, networks, etc.). Listed in table 2 are some of the medium-sized crystal hydrates in which the disordered solvent regions have been characterized. All eight examples were analysed to high resolution using X-rays, while neutron analyses were performed on only the first four structures.

3.3.3. *Large Hydrates.* The majority of the macromolecular hydrates studied have been proteins, of which well over 100 have been solved by X-ray diffraction. Several DNA and RNA structures have also been examined. Most of the more detailed solvent analyses in large hydrates have been done at resolutions of less than 1.5 Å. Below 1.5 Å the solvent positions are better defined with respect to the resolution of density peaks, than for lower resolutions. Extensive reviews of the analysis and implications of water around protein molecules have been given by Finney [104], Edsall and McKenzie [15] and Baker and Hubbard [105]. An outline of some of the protein crystals examined at high resolution, is given in section 6.1.

3.4. *Neutron Diffraction*

This method provides relatively accurate and reliable positions for the hydrogen atoms, the precision of which (average for small structures is 0.01–0.02 Å) is about an order of magnitude better than obtainable from most X-ray analyses. Far fewer hydrates have been analysed with neutrons, than with X-rays. Nevertheless, the available neutron structures provide a reliable source from which detailed information about the bonding and packing characteristics of H-bonded structures can be gained. Several surveys of bonding environments (H-bonding and ion–lone-pair) around water molecules in small neutron structures have been undertaken. Details from the recent survey by Chiari and Ferraris [46] are given in section 2.2.

The number of hydrate structures examined by neutron diffraction is

Table 2. *Medium-sized hydrates in which the solvent has been characterized*

Crystal hydrate	Radiation (X, X-ray; N, neutron)	Resolution (Å)		No of waters	Refs	Comments
		X	N			
Monocarboxylic acid vitamin B_{12}	X, N	~1.0, 1.0		~15	91, 92	About 7 H positions located for the 15 water molecules
α-cyclodextrin	X, N	0.9, ~1.0		6	93	Well ordered; ring systems of H-bonds observed
β-cyclodextrin	X, N	0.9, 0.6		11–12	94, 95	Disordered Os and Hs; systems of 'flip-flop' H-bonds assigned
Coenzyme B_{12}	X, N	0.9, 0.9		14–18	96–99	Extensively disordered, > 140 water and 4 acetone sites assigned, several alternative solvent networks formulated
d(CpG) proflavine	X	0.8		27	100, 101	Observed to be highly structured; 5 pentagonal ring systems assigned
[Phe⁴Val⁶] antamanide	X	0.8		12	102	Several alternative water networks present
Adenyl−3′,5′-uridine, amino-acridine complex	X	~1.0		15	103	Two alternative water networks assigned

continually increasing. In 1968 there were about 20, in 1972 about 40 and at present, over 250. Of these, the majority (greater than 200) are small structures, while the remainder (20 or so) are medium and large structures. In the small systems, very few independent extended networks of water molecules are present containing more than about 2–3 independent water molecules (per asymmetric unit). In ionic hydrates, most of the water molecules are bound to the ions and relatively few participate in water–water H-bonded networks. This also applies to some of the more recently analysed larger ionic hydrates, which contain up to about 10 water molecules. More extensive water networks are present in the medium and larger hydrates, but within these systems there are inherent problems in interpreting the experimental solvent density.

3.4.1. *Small Hydrates.* These provide the most accurate and extensive information about the H-bonding, coordination geometries and short-range order of water molecules that are available. The number of water molecules in each hydrate ranges from one up to about ten per asymmetric unit. The majority of these hydrates are ionic and the electrostatic fields associated with the ions strongly influence the interactions with the surrounding water molecules (dipoles, electron displacement, etc.).

3.4.2. *Medium Hydrates.* Only a few of these structures have been studied by neutron diffraction (four listed in table 2), although a large number of medium-sized hydrates has been examined by X-rays. The first relatively large structure (larger than a small molecule structure) to be studied by neutron diffraction was the monocarboxylic acid derivative of vitamin B_{12} containing 180 atoms [92,106]. The analysis of the solvent regions in this hydrate revealed only about two-thirds of the expected number of water positions and was incomplete despite the high resolution (1.0 Å) of the diffraction data used. The reason for this was probably due to the inaccuracies of the data collected at that time. At present there are only three hydrate structures that are of sufficiently high resolution (less than 1.0 Å) and accuracy to be considered for solvent analysis. These are as follows (see table 2).

α-*cyclodextrin*: essentially this complex is made up of six glucose molecules which are bonded together through the 1–4 oxygen linkages to form a ring. Six water molecules are situated within the crystal (two inside the ring) in well-defined positions that do not appear to be disordered, although one of the hydroxyl groups adjacent to a water position is disordered between two positions. The complete structure is described by Klar *et al.* [93].

β-*cyclodextrin*: this complex is very similar to the α-form except that there are seven glucose units in the ring instead of six. The crystal contains about 11.5 water molecules, distributed over 16 sites most of which are partially occupied [94,95]. The majority of water oxygens and hydrogens and hydroxyl hydrogens are disordered. No details of the water structure (networks, etc.) have as yet been reported.

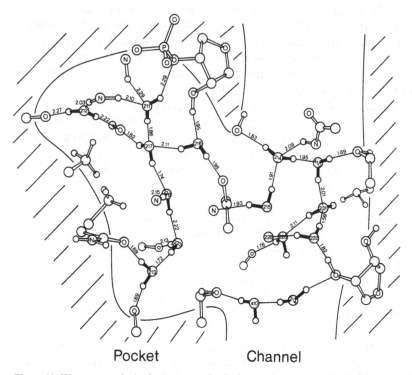

Pocket   Channel

Figure 11. Water networks in the coenzyme B_{12} hydrate structure: the filled bonds represent the water molecules. The solvent distribution in this hydrate is divided into two regions: pocket (left) and channel (right). One of several alternative water networks is shown in each region: network A in pocket and network B in channel. H$\cdots$O bond distances are given in ångströms.

Vitamin B_{12} coenzyme: this complex is essentially composed of a corrin ring (plus side chains), a nucleoside and a nucleotide [107]. The solvent parameters (positional, temperature factors, occupancies) in this system have been extensively refined with one set of neutron data and three different sets of X-ray data. The results of the interpretation of the solvent densities and formulation of water networks were seen to be self-consistent between the four different models, using the standard criteria of H-bonding (O$\cdots$O less than 3.3 Å, H$\cdots$O less than 2.4 Å) and van der Waals contacts (0 = 1.5 Å, H = 1.0 Å). Between 14 and 18 water molecules along with an acetone molecule are present in the solvent regions. The solvent is distributed over two regions – a pocket and a channel (figure 11). The acetone molecule (occupancy $\simeq$ 0.5) is located between the two regions, effectively isolating each of them. When the acetone molecule is absent, the solvent regions are connected to each other. Several different networks of various sizes, containing seven or more water molecules were formulated using the given criteria: the two most ordered networks are shown in figure 11. The solvent in the channel region is seen to be more disordered than in the pocket. Full details of the refinement and analysis of the solvent in coenzyme B_{12} crystals are described elsewhere [96–99].

Of the currently available neutron hydrate structures, the coenzyme B_{12} solvent system gives one of the most detailed experimental descriptions of water structure.

3.4.3. *Large Hydrates*. To date only two protein structures have been analysed by neutron diffraction to resolutions higher than 1.5 Å. These are crambin [108] and hen egg white lysozyme [109], for which complete 1.4 Å resolution data (nominally to 1.2 Å for crambin) have been obtained from crystals deuterated by soaking. The more ordered surface solvent sites correspond quite well with the X-ray positions, but agreement between the more disordered regions is not as good. This may possibly result from the solvent structure being somewhat different in the outer regions or it could be an artefact due to certain sites being assigned to the higher background level noise peaks in these larger systems.

Several neutron analyses of proteins at resolutions of between 1.8 and 2.2 Å have been reported, but very little solvent structural information (hydrogen positions, H-bonded networks, etc.) has been forthcoming, even from the surface of the biomolecules. Further discussion on problems involved in the interpretation of solvent in larger systems is given in section 6.2.

4. Details of Water Geometries in High-Resolution Neutron Structures

Even though we know, in terms of the positions occupied, the actual experimental water structure from crystal diffraction data, the basic structural rationale of why a particular set of positions are occupied is generally not understood in detail. In this section the short-range interactions around individual water molecules are examined in closer detail, in order to assess which may be important in determining the local structure. Some interesting features are evident (section 5) which may help to improve some of the models presently used to simulate water structure in a variety of environments.

The interactions of a given water molecule with those surrounding it can be considered in terms of three components:

(1) $O \cdots O$ interactions, mainly of a repulsive nature;
(2) $H \cdots O$ interactions, attractive or repulsive;
(3) $H \cdots H$ interactions, usually considered repulsive.

However, in the majority of hydrate crystals other interactions are also present (for example in amino acids and other organic hydrates) and the most common specific interactions with the water molecules can be listed as follows:

(1) $O(W) \cdots Y$ where Y is O, N, C or another relatively heavy atom;

(2) O(W)···H where H is attached to either a polar or apolar group;

(3) H(W)···Y where Y is O, N, C or another relatively heavy atom;

(4) H(W)···H where H is attached to either a polar or apolar atom.

The geometries of these structural components have been examined in a number of crystalline solids for which both high-resolution neutron structures were available and the solvent densities were well defined and interpretable. A bibliography of the hydrate crystals analysed is given in the appendix (neutron structures). Included were the ice polymorphs, amino acids, carbohydrate complexes, oxalic acid and similar small organic complexes, some hydrates containing ions (ion–water interactions were not included) and some of the relatively sparse number of medium-sized hydrates: vitamin B_{12} coenzyme and the α- and β-cyclodextrins. These structures contained up to about 18 independent water molecules per crystal asymmetric unit.

4.1. Coordination and Hydrogen Bonding

The H-bonding coordination of the water molecules in *all* the ice polymorph structures is found to be four. In the hydrate structures, excluding water molecules bound to ions, coordinations of three and four are most common with the ratio between these values varying among the different hydrates. Coordinations of three or less tend to occur in regions occupied by a large number of apolar groups such as CH_x. The directionality of the hydrogens of the H-bonds to the lone-pairs region is quite variable: large distortions from the tetrahedral angle are observed.

Two orientational angles, $\theta1$ and $\theta2$, for a water molecule are defined in figure 12. These angles describe the position of an atom X relative to the water position: $\theta1$ is the angle between the bisecting H-O-H vector and the vector from the water oxygen to the position of atom X, while $\theta2$ is the angle between the plane of the water molecule (H-O-H) and a second plane defined by the bisector line and X.

The distribution of H-bonding hydrogens over the lone-pair region, with

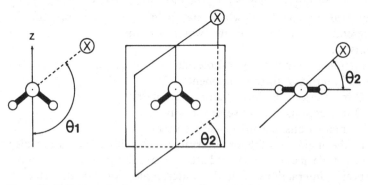

Figure 12. Definition of orientation angles, $\theta1$ and $\theta2$, for a water molecule.

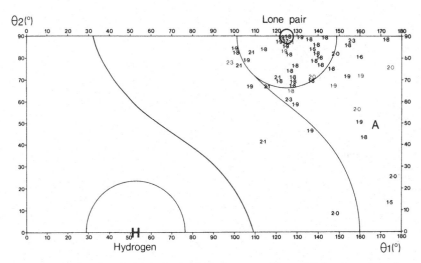

Figure 13. $\theta 1/\theta 2$ orientational angle plot for hydrogen positions of H-bonds made to the lone-pair region (A) of water molecules. Bold numbers are $O(W)\cdots O$ H-bonds (Å) and lighter numbers are $O(W)\cdots N$ H-bonds (Å).

respect to the orientational angles $\theta 1$ and $\theta 2$, is shown in figure 13. Although most of the H-bonds point in the general direction of the lone-pair region (A), they do not appear to be tightly clustered around either the tetrahedral ($\theta 1 = 125°$, $\theta 2 = 90°$) or the trigonal ($\theta 1 = 180°$) positions. This is also seen in the cation distribution plot of figure 8(d) in which the $\theta 1$ orientational angle is defined as $\omega_1 = 180 - \theta 1$ (figure 8(a)). In this plot, hydrogens are found at positions away from the YZ bisecting plane and the tetrahedral position ($\omega_1 = 55°$, $\psi_1(\theta 2) = 90°$).

Figure 14 shows a plot of the $O(W)$–$H\cdots O$ H-bond angles *versus* (a) the $H\cdots O$ distance and (b) the $O\cdots O$ distance for H-bonds made by waters. The geometries follow the same trends as those observed for H-bonds in general: there tends to be a fairly large distribution of individual values for each geometry. The $O\cdots O$ distances range from 2.5 to approximately 3.1 Å; $H\cdots O$ distances from 1.5 to 2.4 Å; the O–$H\cdots O$ angles lie between 120 and 180° (majority 150–180°). No new information about H-bonding is apparent from the larger hydrates. However, two points can be made that relate to non-bonded $O\cdots O$ contacts. First, cut-off lines can be drawn in figure 14, which approximately represent the extreme H-bonding geometries allowed (due to $O\cdots O$ repulsion). These lines may be referred to as the *H-bond bending limit curves*. No geometries are allowed that lie substantially below the curve in the excluded region. Second, some of the longer $O\cdots O$ distances which are greater than 3.1 Å do not appear to correspond to H-bonds: *no* hydrogen atoms between these particular oxygens were observed in the neutron diffraction density maps. As seen in the following sub-section, these

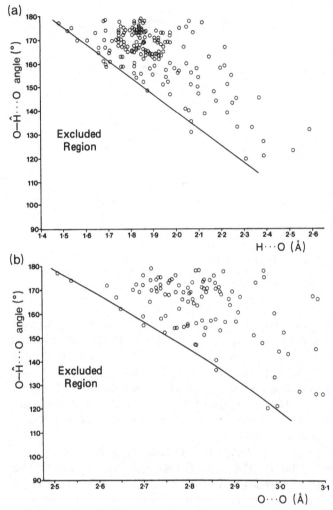

Figure 14. (a) Plot of O–H⋯O angles *versus* H⋯O distances for water H-bonds. (b) Plot of O–H⋯O angles *versus* O⋯O distances for water H-bonds. The full curves represent (approximately) the minimum allowed values.

may be close non-bonded contacts between adjacent waters due to repulsive restrictions of H⋯H and H⋯O non-bonded contacts.

4.2. *O⋯O, C Non-bonded Contacts*

In many of the hydrate structures, some seemingly strange orientations (with respect to conventional wisdom) of the water molecules were observed in relation to both other water molecules and surrounding groups (polar and

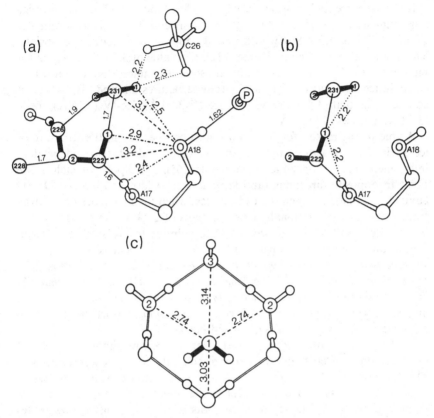

Figure 15. (a) and (b) show close non-bonded contacts around hydroxyl OA18 in coenzyme B_{12}. (c) Close non-bonded O···O contacts in ice-VIII. All distances are given in ångströms.

apolar) within the crystals. Some of the water molecules had close O···O contacts down to about 3.1 Å, yet there were no hydrogen atoms located between them. It could be argued that if the van der Waals radius of an oxygen atom is approximately 1.5 Å (values used in the literature range from 1.4 Å [110] to 1.65 Å [111]), O···O contacts down to 3.0 Å (or 2.8 Å) would be quite acceptable.

The presence of some close O···O contacts can be partially explained by the local H···H repulsive contacts. For instance, in one of the water networks in coenzyme B_{12} crystals (network B) there are 4 or 5 closely packed water molecules adjacent to two hydroxyl groups (figure 15(a)). Two close O···O contacts are made to hydroxyl OA18 by waters W222 and W231. H1 of water 222 cannot orientate itself to form an H-bond to OA18 because of the close H···H contact of H1 to the hydrogen of hydroxyl OA17 (figure 15(b)) and likewise H1 of water 231 cannot form a strong bond to OA18 because of the close H···H contact to H1 of water 222.

In all the ice polymorphs, except ice-VII and ice-VIII, the *average* minimum non-bonded O···O contacts is around 3.5 Å. Thus a van der Waals radius of 1.7–1.8 Å would initially appear to be more appropriate. This point has been cited by Kamb [112]. If a value of 3.5 Å is accepted for minimum O···O contacts, then why are shorter contacts that are known to occur in ice structures and other hydrate structures allowed? Examples of contacts less than 3.5 Å are seen in ice-II (3.24 Å), ice-IV (3.14 Å), ice-V (3.28 Å) and ice-VI (3.34 Å).

A clue to the reason for these variations can be seen in the structure of ice-VIII. This structure (figure 5(b)) is composed of two independent interpenetrating H-bonded ice-Ic lattices [53,65]. The dipoles of each lattice point in opposite directions (anti-ferroelectric). Close non-bonded O···O contacts are made between the two lattices: when the lone pairs point away from each other, exceptionally close contacts of 2.74 Å occur (O1···O2 in figure 15(c)), and when the lone pairs are pointing towards one another in a head-on arrangement, longer contacts of 3.14 Å occur (O1···O3). *This suggests that the repulsive contacts may be variable depending on the relative orientational approaches of the lone pairs and further indicates that the molecular shape of the water molecule in terms of the distribution of electrons may be a significant factor in the short repulsive contacts.* Further supporting evidence for this hypothesis is discussed below.

In the crystal hydrates analysed, the next-nearest neighbour O···O and O···C contacts of less than 3.6 Å around each water molecule were examined. The positions of the neighbouring non-bonded atoms around the water molecules were assessed with respect to the orientational angles $\theta 1$ and $\theta 2$ defined in figure 12. Figure 16 shows plots of the $\theta 1$ against the $\theta 2$ angles of the neighbouring non-bonded atoms (X) for (a) four coordinated water molecules and (b) three coordinated water molecules. The non-bonded contact distances are plotted on one quarter of the $\theta 1/\theta 2$ phase space – the other three are related by the C_{2v} symmetry of the water molecule. These plots show some interesting regularities as follows.

4.2.1. *Four-coordinated (H-bonded) Water Molecules.* This distribution appears as a clear belt of close contacts of less than 3.3 Å for O···O and less than 3.5 Å for O···C, extending from $\theta 1 = 120°$, $\theta 2 = 0°$ to $\theta 1 = 55°$ to $\theta 2 = 90°$ – the middle region C in figure 16(a). In the region over the lone pairs A, contacts are 3.5 Å or greater for both O···O and O···C, while in the area near and between the hydrogens (region B), contacts down to approximately 3.3 Å are observed.

All the observed close non-bonded contacts in the ice polymorph structures lie within region C. For example $\theta 1 = 122°$ and $\theta 2 = 0°$ for the close O···O contact of 2.74 Å in ice-VIII. In ice-II the $\theta 1$ and $\theta 2$ angles of the two water molecules that are in close contact (O···O = 3.24 Å) lie within region C ($\theta 1 = 77.0°$, $\theta 2 = 71.0°$ for water 1 and $\theta 1 = 116.1°$, $\theta 2 = 15.0°$ for water 2).

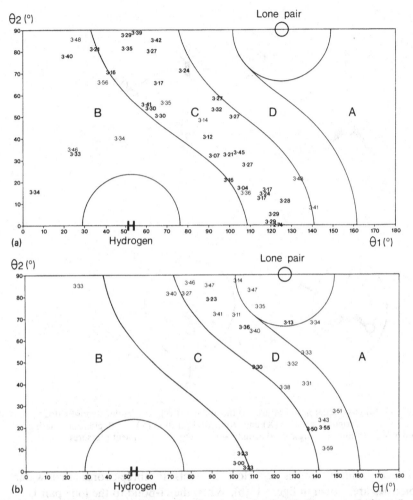

Figure 16. $\theta1/\theta2$ orientational angle plots for non-bonded $O\cdots O$, C contacts (in ångströms) around (a) four coordinated water molecules and (b) three coordinated water molecules. Bold numbers are $O(W)\cdots O$ contacts and lighter numbers are $O(W)\cdots C$ contacts.

In the coenzyme B_{12} hydrate, the corresponding angles for water molecules 222, 226 and 231 with close $O\cdots O$ contacts also occur in region C.

4.2.2. *Three-coordinated Water Molecules.* These water molecules usually form three H-bonds, one by each of the water protons and a third to the lone-pair region with $\theta1$ angles which range from approximately 100 to 180° (trigonal planar). Most H-bonds to the lone pairs are situated at some point between these extreme values. Other factors such as $H\cdots O$ and $H\cdots H$ contacts appear to have a significant influence on the final directionalities of the H-bonds – a point to be discussed in sections 4.3 and 4.4.

(a)

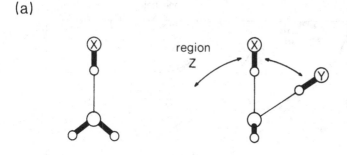

(b)

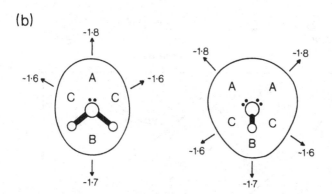

Figure 17. (a) H-bonding positions around the lone-pairs region of a three-coordinated water molecule: trigonal position (X) and tetrahedral position (Y). (b) Regions of differing van-der-Waals radius (in ångströms) around a water oxygen at ambient pressures.

The $\theta 1$ and $\theta 2$ angles for close contacts to three-coordinated water molecules are shown in figure 16(b). When the H-bond to the lone pair is in the trigonal position (i.e. $\theta 1 = 180°$), the $\theta 1$ and $\theta 2$ values for the close $O \cdots O$ and $O \cdots C$ contacts lie mainly within region D: closer to the lone-pair region than for four-coordinated molecules. When the H-bond is near one of the tetrahedral positions ($\theta 1 = 100\text{--}140°$), some of the $\theta 1$ and $\theta 2$ values for the close contacts on the opposite side of the water molecule (region Z in figure 17(a)) fall inside the lone-pair region A.

The above distributions of $\theta 1$ and $\theta 2$ angles, especially for the tetrahedral water molecules, strongly suggest that the van der Waals radius is variable over the water oxygen and is not isotropic. This is exemplified in the ice-VIII structure (figure 15(c)). Three different average minimum contact values may be considered depending on the direction of the close non-bonded contact to the water molecule: (1) region A around the lone-pairs region, minimum contacts of approximately 3.5 Å are allowed; (2) region B around and

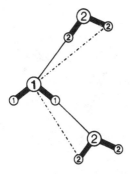

Figure 18. H2···O1 non-bonded contacts (chain lines) between water molecules, with covalent O–H distances regularized to 0.8 Å.

between the hydrogens, minimum contacts down to approximately 3.3 Å are allowed; and (3) region C between the lone pairs and the hydrogens, where minimum contacts down to approximately 3.1 Å are allowed. These areas are shown schematically in figure 17(b). The following van der Waals radii can be approximately assigned: 1.8 Å (region A), 1.7 Å (region B) and 1.6 Å (region C). Thus, the oxygen can be considered to have an asymmetrical repulsive core. Details of the density of electron-cloud overlap from quantum-mechanical calculations may help here in estimating the intermolecular non-bonded contacts between water molecules more accurately.

4.3. H···O Non-Bonded Contacts

In the H-bonding configuration O1–H1···O2–H2, between two water molecules, two types of H···O interactions occur. The first type, H1···O2, constitutes H-bonds which have both attractive (mainly electrostatic) and repulsive components. The second are the H2···O1 interactions between the *remote nearest neighbours* of the H-bond (see figure 18). From a consideration of spherical van-der-Waals radii, contacts of 2.5–2.7 Å between H2 and O1 may be expected. However, in analysing the H2···O1 contacts between water molecules and other adjacent polar groups, several striking regularities are apparent.

One feature of H-bonds that appears at times to be somewhat anomalous, is the assumed correlation between the bending of the O–H···O angles with increasing O···O and H···O distances – see sections 2.4 and 4.1. In figure 6 a regression line (full line) appears to define a correlation. However, in very accurately determined neutron structures (standard deviation of positional parameters less than 0.02 Å), the correlation does not strictly hold, and a large scatter exists. This is clearly seen for water H-bonds plotted in figure 14. Within the hydrate structures, some short H-bonds appear to be more bent (small O–H···O angles) than expected, while some longer bonds

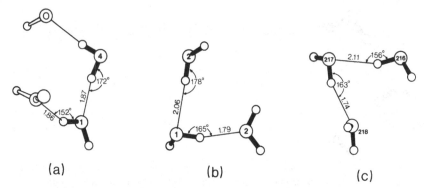

(a)   (b)   (c)

Figure 19. Variations in H-bond bending in (a) α-cyclodextrin, (b) arginine dihydrate and (c) coenzyme B_{12}. H···O distances are in ångströms.

are much straighter than expected. Examples of this are shown in figure 19 for α-cyclodextrin, L.arginine.$2H_2O$ [113] and coenzyme B_{12} hydrate structures. Within these structures it is possible to re-orientate the water molecules in order to increase the smaller H-bond angles; however, *this would present a different structure than that given by the experimental neutron data.*

For a majority of the severely bent short H-bonds, the remote H2···O1 contacts were seen to have minimum distances of around 3.1 Å between the nuclear positions. Further analysis of these interactions involved regularizing all covalent O-H bond distances to 0.8 Å, to represent the electron density over the hydrogens and as a possible centre of interaction. This distance was chosen as a compromise between values of about 0.7–0.9 Å obtained from electron deformation analyses and approximately 0.85 Å obtained in X-ray structural analysis [42,114]. The H2···O1 contacts in all the hydrate structures examined were seen to have minimum values of 3.0–3.1 Å for O1 = a water oxygen, 3.1–3.2 Å for O1 = an uncharged nitrogen and 2.9–3.0 Å when O1 = a hydroxyl, a carbonyl oxygen or part of (or attached to) a charged group.

Figure 20 shows the H2···O1 contacts in (a) α-cyclodextrin for three water molecules, (b) L.arginine dihydrate and (c) the two main water networks (A and B) in coenzyme B_{12} hydrate crystals. The H2···O1 minimum contacts between the waters and the surrounding atoms are all very similar in each of these examples. The water molecules within these structures are orientated such that *the minimum H2···O1 contacts are maintained* around each water molecule *at the expense of distorting the local H-bond geometries*, which are flexible enough in terms of relatively small energy losses to accommodate these contacts. In some cases quite large distortions of the O–H···O angles are observed where the H2···O1 contacts are at their minimum. It does not necessarily follow that bending of H-bonds is entirely due to H2···O1 contacts. The other repulsive interactions of O···O (H-bonds and non-

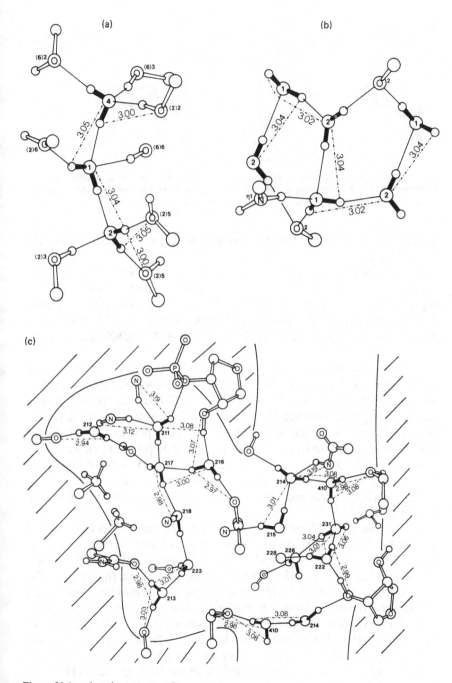

Figure 20 (continued on next page).

(d)

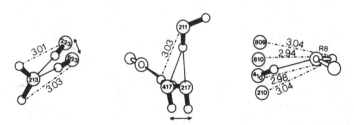

Figure 20. Close H2···O1 contacts (chain lines; O–H = 0.8 Å) in (a) α-cyclodextrin, (b) arginine dihydrate, (c) coenzyme B_{12} and (d) of local disorder in coenzyme B_{12}. All distances given in ångströms. All remaining HS···O1 contacts are greater than 3.1 Å.

bonded) contacts and H···H contacts may also result in distortions of the local H-bonds.

Areas of local solvent disorder also appear to comply with these proposed minimum H2···O1 contacts. For example, in solvent networks of coenzyme B_{12}, the extreme positions of the local disorder of several water molecules depend on restrictions imposed by the minimum limits of the H2···O1 contacts: water regions 223/623, 217/417 and 410/210/809/810 in figure 20(d).

Further supporting data for the minimum H2···O1 contacts in H-bonding comes from the ice phases. In ice-Ih all water molecules are involved in six-membered rings (figure 21(a)). The H2···O1 contacts are all approximately 3.1 Å in the tetrahedral structure, and it is possible that the O···O H-bond distance could shorten slightly. If the water positions in this ice are slightly disordered – analysis of recent neutron data [115,116] indicates that this is probably the case – then the unusually long O–D bond lengths of 1.01 Å observed (compared with an average of 0.98 Å in the other ice polymorphs [82]) may be accounted for as an artefact of previously undetected oxygen disorder.

All the minimum H2···O1 contacts in the high-pressure ices are around 2.9–3.0 Å. The ice-II structure is composed of columns of hexagonal rings which are connected through their outer H-bonds both in the vertical direction of the column and horizontally to the next column (figure 2(c)). The coordinations of the H-bonds involved in the connecting bonds are appreciably distorted from tetrahedrality and some of the minimum H2···O1 contacts of the connecting H-bonds are down to 2.96–2.97 Å. In ice-V and -IX(III), the molecules are involved in five-membered rings. An accurate neutron structure exists for ice-IX where hydrogens are present at two levels of occupancies, approximately 0.96 and 0.04. Minimum H2···O1 contacts of 2.97–3.10 Å occur (figure 21(b)). In ice-V H2···O1 contacts of the five-membered rings vary from 2.95 to 3.08 Å. Four-membered rings are

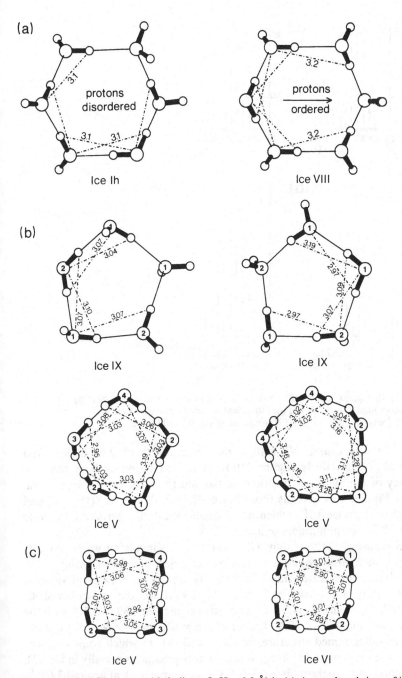

Figure 21. H2···O1 contacts (chain lines; O–H = 0.8 Å) in (a) six-membered rings of ices-Ih and -VIII; (b) five-membered rings of ices-IX and -V; (c) four-membered rings of ices-V and -VI. All distances given in ångströms.

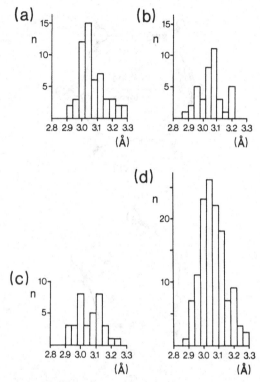

Figure 22. Histogram of short H2···O1 contacts. (a) water–water in hydrates; (b) water–water in high-pressure ices (including less accurate structures); and (c) water–hydroxyl in hydrates; and (d) composite of (a), (b) and (c).

present in ices-V and -VI with minimum contacts of 2.92–3.08 Å and 2.89–3.01 Å respectively (figure 21(c)). It should be noted here that the accuracy of the hydrogen positions in the neutron structures of ices-V and -VI [63,53] is not as good as those in ices-Ih [55], -II [58], -IX(III) [60] and -VIII [53,65]. In ice-VIII, which has a similar sub-lattice structure to ice-Ic (section 2.1), the minimum contact is 3.2 Å.

Histograms of the minimum H2···O1 contacts involving water oxygens found in the hydrates and ice structures examined, are shown in figure 22. The hydrate water–water H2···O1 contacts tend to have a cut-off value at approximately 3.0 Å (sharp decrease to 2.9 Å). In the ice polymorphs examined, the minimum H2···O1 contacts range from 2.89 to 3.2 Å with the majority around 2.96–3.05 Å. The shorter contacts mainly occur in the less accurately determined structures of ice-V and -VI, in which some disorder of both the oxygens and hydrogens is probably present (especially in ice-VI). The combined histogram (figure 22(d)) shows a large peak at around 3.04 Å.

In general there appears to be a very much larger spread in the H-bond geometries than in the minimum H2···O1 contacts. The O···O and H···O

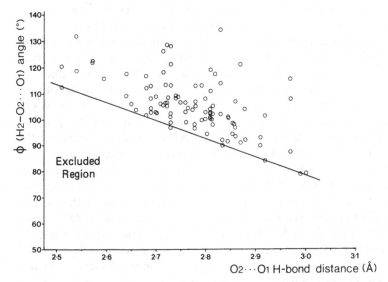

Figure 23. Plot of H2–O2···O1 angles (ϕ) *versus* O2···O1 H-bond distances for waters involved in H-bonding. The full line represents the minimum allowed values.

H-bond distances vary from 2.5 to 3.2 Å and 1.5 to 2.4 Å respectively, *but the minimum H2···O1 contacts vary over a much smaller range*: 2.9–3.1 Å. This strongly suggests that the H2···O1 contacts are relevant to the orientational control of the H-bonding geometries, much more than the conventionally stressed orientational dependence of the attractive H-bonding interactions.

When the O2···O1 H-bond distances are plotted against ϕ, the angle H2–O2···O1 (in the H-bonded conformation H2–O2···H1–O1), ϕ, is seen to increase as the O2···O1 H-bond shortens: see figure 23. The full line in the figure approximately represents the *minimum values* of ϕ for given H2–O2···O1 geometries that are permitted without violating the minimum H2···O1 contacts of approximately 3.0 Å. This line – the *H2···O1 contact limits curve* – approximately separates the H2–O2···O1 conformations into two regions: an area of excluded geometries which lies below the curve and an allowed area above. The repulsive characteristics of the H2···O1 interactions, that is how hard or soft they may be, are not known exactly. The use of a sharp cut-off curve (i.e. a hard-sphere model) is an over-simplification, however, it is easily seen that conformations which lie substantially below the limiting curve are not allowed.

The physical nature of the H2···O1 interactions is difficult to rationalize in terms of classical van der Waals contacts. Taking the commonly accepted radii of approximately 1.5 Å for oxygen and 1.0 Å or even 1.2 Å for hydrogen, only gives a maximum of 2.7 Å which is significantly smaller than the H2···O1 contacts. *One plausible explanation of these longer contact distances, using the concept of van der Waals radii, is that the radius around the hydrogen is variable* – probably shortened in the forward O-H direction. It is well known that the electron cloud of a polar hydrogen is pulled toward the electronegative atom to which it is bonded, thus making the electron density over the hydrogen somewhat aspherical. The redistribution of electron density over certain atoms often results in significant changes of their van der Waals radii. An example of this is ionization: the van der Waals radius of an uncharged sodium atom is about 1.9 Å and decreases to approximately 1.0 Å on the loss of an electron, while for chlorine the opposite effect takes place, the radius increases from approximately 1.0 Å to 1.8 Å on ionization to Cl.⁻ Similar effects probably occur for the van der Waals radii of polar water hydrogens with respect to a shift in electron density from the hydrogen to the oxygen. In the region towards the oxygen, a radius of 1.4–1.5 Å may be valid, while in the direction of the O-H vector it may have a value of 0.9 Å or less (figure 25(a)). This suggests the relevance of polarizability to water–water interactions, but in terms of perturbing repulsive interactions rather than the usual framework of enhancing attractive interactions, in which polarization has normally been discussed [79, 80].

A clearer physical explanation, if it is possible, of the remote H2···O1 contacts (may be in terms of O-H bond interactions or electron-cloud overlap), will have to await detailed quantum-mechanical analyses. *What is evident* though, is that some sort of interaction that keeps the remote H2 hydrogens away from the O1 oxygens seems to be present and is relevant to the resulting orientational structure.

4.4. *H···H Non-bonded Contacts*

Two types of H···H contacts are present in most hydrate crystals. These are (a) contacts between a water molecule and other waters or local polar groups (OH, NH etc) of the c/molecule and (b) contacts between a water molecule and apolar groups (CH_x) of the c/molecule:

4.4.1. Water–Water H···H Contacts. The average minimum values for these contacts in the hydrates and ices examined, lie between 2.3–2.4 Å for distances up to 2.6 Å. In the ice polymorphs the shortest H···H contacts range from 2.24 Å (ice-IX) to 2.39 Å (ice-VIII), while in the crystal hydrates minimum values of 2.05 Å to 2.5 Å are present. As in the case of the H2···O1 non-bonded contacts, there is some evidence that the van der Waals radii of water hydrogens may be variable, having some angular dependence. Figure 24(a) shows the H···H water–water contact distances as a function of the sum χ of the angles $\alpha + \beta$ subtended at each of the two hydrogens (see inset in figure 24(a)). As χ increases, the H···H contacts tend to decrease and a line representing the limiting values of χ and H···H can be drawn (values that lie significantly below this line do not appear to be acceptable). In the region between H···H = 2.4 and 2.7 Å and χ less than 180° there are fewer

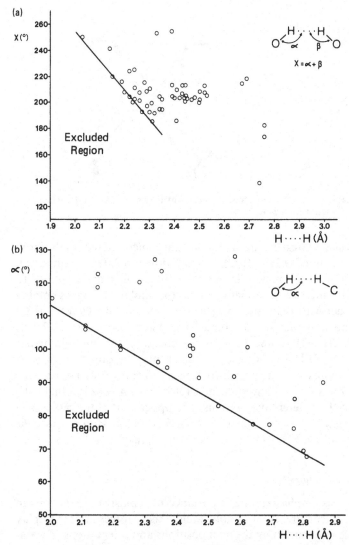

Figure 24. (a) Plot of water–water H···H contacts (internuclear) *versus* χ, the sum of the angles α and β subtended at the hydrogen atoms (see inset). (b) Plot of water-apolar H···H contacts (internuclear) *versus* α, the O(W)–H···H(C) angle subtended at the water hydrogens.

points. This appears to be due to the dominance in this region of H2···O1 remote contacts, which essentially prevent many water orientations with α and β angles between 50 and 140°.

4.4.2. *Water–Apolar H···H Contacts.*
Minimum values of between 2.0 and 2.7 Å are observed. The H···H water–apolar contact distances are plotted against their respective O–H···H(C) angles (α) in figure 24(b). Again a line

(a)   (b)

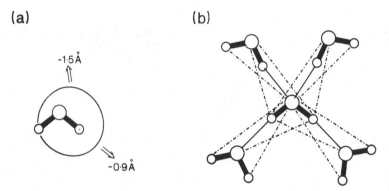

Figure 25. (a) Variable van-der-Waals radius for a water hydrogen. (b) H2···O1 contacts (chain lines) around a tetrahedral water molecule.

can be drawn which represents the minimum values encountered. In the more head-on arrangement of O-H ··· H(C), contacts down to 2.0 Å are observed, but at smaller angles the minimum H···H contacts increase from 2.0 to 2.7 Å. Part of this increase is probably due to the apolar hydrogens coming into closer contact with the water oxygens since most of the O(W)··· H(C) contacts near the minimum H···H contact line (figure 24(b)) are 2.6–2.7 Å.

The minimum H···H contacts appear to depend on the mutual angular orientations of the O-H vectors and this effect may be due to a variable van-der-Waals radius for the water hydrogens. Values ranging between approximately 1.0 Å or less (α greater than 150°) and 1.4 Å (α less than 60°) may be valid for the different orientations. The upper limit of the radius is in line with the implied value of approximately 1.5 Å for the H-O direction (α less than 60°) from the H2···O1 contacts (figure 25(a)).

4.5. *Summarizing Remarks*

The above analysis emphasizes the importance of repulsive interactions in H-bonded water structure. As in simple liquids, these forces appear to play a significant role in determining the final positions and orientations of water molecules in a particular structure. Several components of the repulsive interactions have been identified, which to a first approximation, can be rationalized in terms of assigning anisotropic repulsive cores to both the oxygen and hydrogen atoms.

5. Implications of Repulsive Limitations in Water Structures

In this section, the role and implications of the repulsive interactions within the structures of the different water phases (hydrates, ices, liquid, apolar contacts) are examined. First, we outline the main factors that appear to control the three-dimensional structure.

5.1. *Basis of Structure*

5.1.1. *Hydrogen-Bonding.* Hydrogen-bonds are undoubtedly the predominant component of the attractive interactions between water molecules. They are strongly directional in that there is a tendency to form linear O–H $\cdots$ O H-bonds, but not necessarily collinear. The energy losses in bending an H-bond by up to 20–30° for weak and medium strength bonds are not very great, approximately 0.1–0.3 kcal [117], thus angles down to 150° are common. It is well known that as the O $\cdots$ O and the H $\cdots$ O H-bond distances increase, the O-H $\cdots$ O angles *can*, if necessary, become more significantly bent (figures 6 and 14) and the limits of H-bond bending depend on the repulsion between the two electronegative atoms (oxygens for water). The extreme bending limits, *H-bond bending limits curve*, are shown as broken and full lines in figure 6 and 14 respectively. It is probably more meaningful to look at the repulsive limits of H-bond bending instead of the usually assumed correlation between the H-bond angle and H $\cdots$ Y distance (the former decrease as the latter increase: full line in figure 6). The limiting curve divides the H-bond geometries into allowed and excluded conformations. The actual angles and distances an H-bond may have, depend on the local packing arrangements determined mainly by short-range non-bonded contacts (see below). The ranges of the water H-bond geometries observed in ice and hydrate structures are: O $\cdots$ O = 2.5–3.2 Å, H $\cdots$ O = 1.5–2.4 Å and O-H $\cdots$ O angles = 130–180°.

5.1.2. *Repulsive Limitations.* As we found in section 4, the minimum repulsive contacts around a water appear to be variable depending on the type of interaction (O $\cdots$ O, H $\cdots$ O, H $\cdots$ H) and the region of the molecule (e.g. lone pair or hydrogen). Four main different short-range repulsive components can be identified, and these are:

(RR1) O $\cdots$ O repulsion of H-bonds;
(RR2) O $\cdots$ O repulsion of next-nearest neighbours;
(RR3) H2 $\cdots$ O1 remote nearest neighbours (repulsive); and
(RR4) H $\cdots$ H repulsive contacts.
The following characteristics of these contacts are evident:

(RR1): H-bond angles tend, on energetic grounds, to be linear unless there are other local packing/repulsive forces that would consequently be highly strained. In this case the O-H $\cdots$ O H-bond angles can bend to within the limits allowed by the H-bond O $\cdots$ O repulsions (*H-bond bending limits curve*), to relieve surroundings short-range forces.

(RR2): The water oxygens appear to have an asymmetrical repulsive core depending on the orientation of the water molecule. Three regions may be assigned with the following van der Waals radii: approximately 1.8 Å over lone pairs, approximately 1.7 Å between the hydrogens and approximately 1.6 Å between the lone pairs and the hydrogens.

(RR3): H2 $\cdots$ O1 contacts have a minimum value of approximately 3.0 Å

(for O-H fixed at 0.8 Å), the majority are around 3.1 Å or greater. The exact nature of these minimum contacts is not known, but they are probably due to distortion of the electron cloud over the hydrogen.

(*RR4*): H $\cdots$ H contacts have a minimum value of approximately 2.1 Å, the majority are around 2.3–2.4 Å. The minima are usually observed in regions where the O-H $\cdots$ H-O orientations tend to be more 'head-on'.

The angular correlations of both the H2$\cdots$O1 and H$\cdots$H contacts strongly suggest that the van der Waals radius around water hydrogens (and probably other polar molecules) is variable depending on the H2-O2$\cdots$O1 and O-H$\cdots$H angles. To a first approximation, a radius between the limits of 0.9 Å (or less) in the direction of the O-H vector and 1.5 Å in the H-O direction appears to fulfil many of the above observations.

The minimum contacts for RR3 and RR4 are in the order of about 0.1–0.2 Å shorter than the average values observed. This is also seen to occur for the shortest O$\cdots$O contacts of RR2 (3.1 Å, average = 3.3 Å). Smaller ranges in the variation of the repulsive contacts are observed, from 0.1 to 0.3 Å, compared with the attractive H-bond interactions which may vary by up to approximately 1.0 Å (H$\cdots$O = 1.5–2.4 Å).

When compared with the apparently more dominant interactions of electrostatics and H-bonding, many of the close non-bonded contacts observed in structural analyses of H-bonded substances (experimental and theoretical) are not usually considered to be very significant, unless they violate the *assumed* van der Waals contacts. In relatively simple systems (e.g. single molecular centres), the short-range repulsive contacts are seen to be a dominant factor in determining the final positions of the individual atoms within a structure [28–33]. On making small structural changes, the repulsive interactions (and their associated high energies) vary much more rapidly than longer range attractive forces. In known water structures, such as ice-Ih, each water has four H-bonded tetrahedrally coordinated neighbours and for distances less than 3.5 Å from a central water (H1–O1–H1), there are:

4 H$\cdots$O   H-bonding interactions (approximately 1.8 Å);
4 O$\cdots$O   repulsive interactions of the H-bonds (approximately 2.8 Å);
8 H$\cdots$H   non-bonded contacts (approximately 2.3 Å); and
12 H2$\cdots$O1   remote non-bonded contacts (approximately 3.1 Å, see figure 25(b)).

Although the relative energy contributions are balanced in favour of the attractive interactions (electrostatic, dispersion, induction, etc.), the number of short-range non-bonded contacts outnumber the H-bonding contacts, 24:4. This ratio may increase somewhat when additional 'intrusive' O$\cdots$O next-nearest-neighbour contacts of 3.0–3.6 Å (e.g. in high-pressure ices) are also included. The collective number of non-bonded short-range contacts play an important part in determining the final positions and orientations of the water molecules in a structure. This is particularly the case for the hydrogen atoms which participate in H-bonds. Some of the implications of the short-range interactions are discussed in the following section.

5.1.3. *Factors Controlling the Short-range Structure.* The basis of water structure appears to depend on two main underlying components: (1) H-bonding, and (2) maintenance of the minimum H2$\cdots$O1 contacts. The second component appears to be a dominant factor in determining the actual geometrical arrangements that the former may have (H-bond distances and angles). The non-bonded O$\cdots$O and H$\cdots$H contacts also play important roles, largely in regions where the structures adopted are highly distorted from tetrahedrality, or where three-coordinated water molecules are present.

To go to a more compact water structure than the most open one of ice-Ih (density = 0.92 g cm^{-3}) which is fully tetrahedral, there has to be a reduction in the next-nearest-neighbour O$\cdots$O distances. There are two ways of doing this: (1) by decreasing some of the O$\cdots$O$\cdots$O H-bond angles or (2) by forming interpenetrating H-bonded networks. The latter method gives rise to some very short O$\cdots$O repulsive contacts in order to retain H-bonding O$\cdots$O distances below approximately 3.2 Å, and is only really viable at higher pressures (ices-VI, -VII and -VIII) where close non-bonded O$\cdots$O contacts are sustained. The first method appears to be the main way of forming more closely packed water structures at lower pressures. When the O$\cdots$O$\cdots$O angles become smaller, some of the minimum H2$\cdots$O1 contacts are substantially decreased (less than 2.9 Å) violating the minimum acceptable value of approximately 3.0 Å.

The only way the O$\cdots$O$\cdots$O angles can be significantly decreased is by increasing one or both of the O$\cdots$O H-bond lengths so that the minimum H2$\cdots$O1 contacts are maintained. This is illustrated schematically in figure 26(a), where for simplicity, the hydrogens are assumed to lie on the O$\cdots$O H-bond vector. Oxygens O1, O2 and O3 initially form two H-bonds of approximately 2.75 Å, in the standard tetrahedral arrangement: the O1$\cdots$O2$\cdots$O3 angle is approximately 109°. The H2$\cdots$O1 distance is about 3.1 Å (O–H = 0.8 Å). When O3 moves to the O4 position to decrease the O1$\cdots$O2$\cdots$O3 angle, H2 consequently moves to H3 (assuming retained H-bond linearity). Then the H3$\cdots$O1 distance become less than the acceptable minimum of approximately 3.0 Å (RR3). Thus, the O1$\cdots$O2 H-bond must lengthen by moving O1 to, say, the O5 position in order to relieve the H3$\cdots$O1 repulsive strain.

Where necessary, the hydrogen bonds can bend as long as their H-bonding geometries are within acceptable limits (above the *H-bond bending limits curve* in figure 14). Figure 26(b) shows schematically the inclusion of H-bond bending in the decrement of the O$\cdots$O$\cdots$O H-bond angles. For O$\cdots$O H-bonds of 2.75 Å, the O–H$\cdots$O angles can bend by about 20° (figures 6 and 14), hence O3 may move to the O4 position without the necessity of moving H2, and the O1$\cdots$O2$\cdots$O3 angle is decreased to about 100°. A further decrease in this angle may only take place if either of the O1$\cdots$O2 or O2$\cdots$O4 H-bond lengths are increased to maintain the minimum H2$\cdots$O1 contact (as in figure 26(a)).

At relatively small O$\cdots$O$\cdots$O angles, the lengthening of the O2$\cdots$O1

(a)

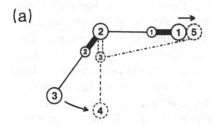

(b)

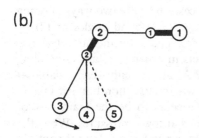

Figure 26. (a) Schematic diagram of decrease in O···O···O H-bond angles and maintenance of minimum H2···O1 contacts. (b) O···O···O angle decrease including the effect of H-bond bending.

H-bonds leads to weaker or non-bonded interactions (greater than 3.1 Å). If a significant number of unmade H-bonds are present within the bulk phase, then the overall structure would tend to be less stable than a fully bonded assembly. Therefore, in the interests of maintaining the maximum possible number of H-bonds, it would appear to be energetically more favourable to form either partially or totally interpenetrating H-bond lattices that are close to tetrahedrality.

The shortest possible H-bonds that may be involved in tetrahedrally coordinated structures (O···O···O = 109.5°) and still maintain the minimum H2···O1 contacts, are about 2.7 Å. For short H-bonds down to approximately 2.5 Å, H2···O1 contact restrictions are placed on the acceptor water molecule which has to adopt a conformation that tends towards being trigonal planar. This has to be done in order to relieve the short H2 (acceptor) to O1 (donor) contacts (less than 3.0 Å). Thus, acceptor waters that participate in short H-bonds tend to be mainly three-coordinated. Very short H-bonds of less than 2.5 Å appear to have more covalent character and the nature of the H-bond interaction is somewhat different with the covalent O-H bond of O-H···O increasing significantly as H···O decreases [46]. As yet, the characteristics of the local short-range geometries of these very strong H-bonds have not been reported. Where longer O···O H-bonds of 2.8–3.2 Å are allowed, significant distortions from the tetrahedral structure are possible.

At these longer distances the O···O···O H-bond angles are allowed to bend as long as they comply with the minimum H2···O1 contacts (RR3) and the H-bond bending limits curve (RR1).

Hence, in order for water molecules at normal pressures to pack more closely, a proportion of the H-bonds *must* either lengthen or shorten their O···O distances. Shortening of H-bonds allows fewer bonds to be made at the expense of increasing the number of non-bonded O···O contacts. On the other hand, increasing the H-bond lengths does not restrict the maximum number of bonds that can be made, and at the same time allows non-bonded water molecules to interpenetrate between the larger distorted O···O···O H-bond angles.

All four of the different short-range repulsive contacts between the atoms in a water structure (RR1-4) must comply with the minimum values allowed for particular water geometries. The repulsive correlations place some very fine restraints and restrictions on the possible structures that an assembly of water molecules may have, both in terms of mutual intermolecular geometries and volume exclusion (as in simple liquids). By using these criteria, the final positions of the atoms in a water structure can be explained in terms of a general minimization of the four repulsive interactions and a maximization of the H-bonding interactions. This may be done initially with respect to the geometries: by maximizing the four repulsive contacts along with the O-H···O H-bond angles. A fuller analysis requires the use of computer simulation techniques incorporating the above repulsive characteristics either into the potential or as restraints.

One interesting feature of water structure that emerges from the above discussions is that the minimum H2···O1 contacts are seen, to some extent, to enforce a tetrahedrally coordinated structure. This is especially the case in an environment which is highly symmetric such as in a crystal, where it is advantageous in terms of the energetics to form as many short H-bonds as possible, but at the same time minimizing the H2···O1 and H···H repulsive interactions. As pointed out above, the local geometry that satisfies these criteria is one in which there are four H-bonds of lengths in the order of 2.7 Å in a tetrahedral arrangement. Shorter H-bonds lead to a loss in number, while longer ones allow distortions but losses only in bond strength. Thus, it would appear that the rather open structure in hexagonal ice-I is a case where all the water–water interactions are optimized in terms of maximizing the number and strengths of H-bonds and minimizing all the repulsive contacts.

The H2···O1 and H···H contacts influence the packing efficiency (H-bond lengths etc.) within regular ring structures. Five-membered rings (*not necessarily planar*) represent the smallest closed-loop structures allowed, in which reasonably strong H-bonds (2.7–2.8 Å) are preserved, but the H2···O1 contacts are at or very close to their minimum values, approximately 3.0 Å. Examples of such five-membered rings are seen in ices-III(-IX) and -V (see figure 21(b)). For rings containing fewer than five water molecules, some of the O···O H-bond lengths must increase to maintain the minimum H2···O1 contacts, becoming less stable. In addition to this, the H-bond coordination around the individual waters tends to become very distorted from tetrahedrality (O···O···O angles down to 70–80°). In the clathrate hydrate structures five-membered rings are very common [118,119]. They are composed of polyhedral water structures containing cavities of differing sizes which may be occupied and stabilized by a variety of relatively inert 'guest' molecules. Interactions between the water molecules and the enclosed species are usually of van der Waals nature. The water molecules form a cage-like structure around the guest molecule, and one of the most efficient ways of doing this, with respect to (a) forming as many H-bonds as possible and (b) maintaining the minimum H2···O1 contacts, is by forming a large number of five-membered rings.

5.2. Orientations and Packing of Water Molecules

The regularities found in hydrate and ice structures discussed above suggest that the final orientation of a water molecule in an equilibrium structure can be determined from the four main repulsive contacts. The main contacts involved are the O···O H-bond repulsions and the H2···O1 remote neighbour contacts. We now discuss the operation of these restraints in various water involved structures.

5.2.1. Ice Polymorphs.

Ices-Ih and -Ic appear to be the only forms of water structure in which all four of the main repulsive interactions are above their minimum contact distances and the H-bond interactions are fully maximized in terms of the average positions occupied as deduced from diffraction. In the equilibrium structure, each water forms four relatively strong H-bonds, O-H···O angles of approximately 180° and O···O = 2.76 Å, which is close to the shortest distance possible without violating the minimum H2···O1 remote contacts. The H2...O1 contacts are all approximately 3.1 Å when calculated from a tetrahedral network (and O-H = 0.8 Å). These distances along with the H···H (internuclear) minimum contacts of 2.29 Å are longer than the average minimum H2···O1 and H···H contacts of 3.0 and 2.1 Å respectively. The O···O next-nearest-neighbour contacts of 4.5 Å are much greater than their minimum contacts of 3.1–3.6 Å. Small individual and collective movements of water molecules within the minimum contacts are possible and these may well explain the apparently abnormal O-H bond lengths of 1.01 Å observed in ice-Ih by neutron diffraction.

The ice phase diagram for ice I to ice-IX (pressures up to approximately 30 kbar) can be explained tentatively in terms of O···O next-nearest neighbour and H2···O1 contacts. The complete pressure–temperature phase space for the known ice structures is shown in figure 1. The phase space

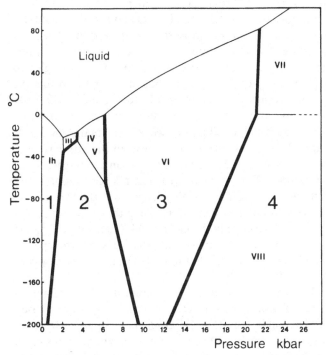

Figure 27. Division of the pressure–temperature ice phase diagram into four regions with respect to O···O and H2···O1 repulsive limitations.

can be roughly divided into four main regions (figure 27) with respect to repulsive contacts that are fully strained within a particular ice structure, preventing it from being further compressed without a complete structural change to relieve the repulsive limitations. The regions and interactions strained within each (at the right-hand boundaries) are:

Region 1, H2···O1 contacts ⎫
Region 2, O···O contacts   ⎬ for a single lattice of H-bonds,
            H2···O1 contacts ⎭

Region 3, H2···O1 contacts ⎫
Region 4, O···O contacts   ⎬ for two independent lattices of H-bonds,
            H2···O1 contacts. ⎭

Region 1: Ice-Ih,-Ic and -III. H2···O1 minimum contacts are 3.0–3.1 Å. O···O next-nearest neighbours in ice-Ih and -Ic are 4.5 Å and greater than 3.45 Å in ice-III (-IX). On increasing the pressure, the H2···O1 contacts cannot be substantially reduced, thus the only alternative is to reduce the O···O non-bonded contacts and hence fill in some of the empty space in the rather open single H-bonded lattice.

Region 2: Ice-II, -IV and -V. These ice structures are composed of single H-bonded lattices in which some very close O···O next-nearest-neighbour contacts of between 3.14 and 3.3 Å occur between water molecules depending on their mutual orientations (section 4.2). The O···O and H2···O1 contacts are at their minimum limits. When the pressure is increased, the structures can use one or both of the following options to relieve the repulsive strains at the phase transition: (1) Decrease some of the O···O···O angles to increase the packing efficiency of the oxygens, but in order to do this the O-H···O H-bond lengths have to be *increased* to maintain H2···O1 minimum contacts. This situation occurs in several of the small and medium hydrate structures. However, the possibility of weaker H-bonds, 2.8–3.2 Å, has to be considered with respect to the balance between the attractive and repulsive interactions: can a decrease in the former sustain an increase in the latter at increased pressures? For symmetry reasons, a large proportion of the H-bonds in crystalline ices would probably lengthen if the O···O···O angles were to decrease substantially, hence decreasing the attractive energy contribution. (2) To maintain relatively strong H-bonds at around 2.8 Å, and maintain minimum H2···O1 contacts, another alternative is to form separate independent H-bonded units or lattices that interpenetrate in such a way to minimize repulsive contacts.

For the transition between regions 2 and 3, the second choice would appear to be energetically more favourable in the case of a pure crystalline phase.

Region 3: Ice-VI. The minimum H2···O1 contacts are 2.9–3.0 Å within this self-clathrate structure. The O···O next-nearest-neighbour contacts are all around 3.4 Å which is just above the average of the three different contact distances around a water oxygen (figure 17(b)). The H2···O1 contacts are highly strained and a further collapse of the structure (with the retention of standard H-bonds) can take place with respect to reducing the slacker O···O contacts. This may be achieved if the two independent H-bonded lattices form a structure in which they are completely interdigitating.

Region 4: Ice-VII and -VIII. These polymorphs are composed of two completely interpenetrating ice-Ic-type sub-lattices and both the O···O and H2···O1 non-bonded contacts in region 4 appear to be at their limits for sustaining the normal type of H-bond (2.5–3.2 Å). In the ordered ice-VIII phase, the H-bond lengths are longer (2.88 Å) than in most of the other ice polymorphs. Within this structure, there are several close H2···O1 contacts between O-H groups which form a cyclic arrangement around the very close O···O non-bonded contact of 2.74 Å (see figure 5(a)). The water molecules are orientated such that the O2–H2···O1 angles are approximately 90° and using an angular-dependent van-der-Waals radius for hydrogen, a value of approximately 1.2 Å may be reasonable in this direction. Assuming the van der Waals radius of the oxygen is about 1.4 Å (half the O···O contact), the shortest H2···O1 contact expected would be about 2.6 Å, which is similar to the observed value of 2.56 Å. Further collapse of the ice structure would probably significantly alter the electronic structure of the waters (ice-X), since all of the eight close $O \cdots O$ contacts in both ices VII and VIII are less than 3.1 Å, the minimum contact observed for the majority of water oxygens.

5.2.2. Hydrates. One striking observation in these structures is that sometimes a water or a hydroxyl group (of the c/molecule) is *partially* surrounded, usually on one side, by several water molecules that do *not* form H-bonds to the partly enclosed species. The reason for this stems from the fact that the local $H2 \cdots O1$ and $H \cdots H$ close repulsive contacts of the surrounding water molecules are at their minimum limits preventing them from re-orientating and forming H-bonds to the central water or polar group. Figure 15(a) shows an example of this: the water molecules surrounding the OA18 hydroxyl group cannot rotate and form H-bonds to OA18 without violating the minimum $H2 \cdots O1$ and $H \cdots H$ contacts. Thus, partial water structures, which may be termed 'partial (clathrate-like) cages' or 'minimized or optimized repulsive structures' appear to be present. Complete interpenetrating 'self'-clathrate structures exist in the high-pressure ice-VI, -VII and -VIII. Partial cages and similar packing arrangements may occur in places where the local repulsive interactions would otherwise be severely strained, for instance, in trying to form extremely distorted H-bonds.

When water molecules are packed in such a way that only non-bonded interactions exist between the groups in question (as in the 'partial cages') then the $O \cdots O$, C next-nearest-neighbour contacts play a significant role. The actual contact distances depend on the mutual orientations of the water molecules (asymmetric repulsive cores) and their relative positioning with respect to other groups (e.g. CH_x). Substantially different water $O \cdots O$ contacts ranging from 3.1 to 3.6 Å or more are possible.

5.3. *Liquid*

As described in section 2.3, the composite neutron *PCF* can be separated into the partial *PCF*s: $g_{OO}(r)$ (X-ray $g_{OO}(r)$ more reliable), $g_{OH}(r)$ and $g_{HH}(r)$ (figure 10). However, difficulties arise in trying to sort out which contributions come from H-bonded, and which come from non-bonded neighbours. The only interactions that can be clearly separated from other significant contributions, are the comparatively short ones from the near neighbours. These are essentially the first peaks in the three partials: $O \cdots O$ H-bonds at 2.85 Å (X-ray), $H \cdots O$ H-bonds at approximately 1.9 Å and the $H \cdots H$ repulsions at approximately 2.4 Å. Each of these peaks is spread out and for distances at their upper ends, it is not known which atom pairs are immediate neighbours (H-bonds) and which are next-nearest neighbours. For example, the X-ray $g_{OO}(r)$ first peak ranges from 2.5 to around 3.3 Å and at more than 3.1 Å, some distances could be weak H-bonds and others short non-bonded contacts. The second peak in the $g_{OO}(r)$ partial occurs at approximately 4.5 Å which is the next-nearest-neighbour distance in the tetrahedral ice-Ih and -Ic structures. The presence and height of this peak indicates that there is significant tetrahedral-'*like*' structure within liquid water, but there is a significant dispersion in the $O \cdots O \cdots O$ angles about the tetrahedral value of 109.5°. In hydrate structures they range from approximately 60 to 150° and a similar range may well occur in the liquid.

The first peak of each partial essentially corresponds to the average geometries observed in hydrate crystals for H-bonds and short contacts: $O \cdots O = 2.81$ Å [46], $H \cdots O = 1.86$ Å [46] and $H \cdots H \simeq 2.4$ Å. More significant information about the longer first peak $O \cdots O$ contacts (greater than 3.1 Å), and second peaks may be obtained from values observed in hydrates (section 4.2). For instance, several of the longer $O \cdots O$ first-peak distances are seen to be short non-bonded contacts around a particular water. This is due to some of the surrounding water molecules being unable to form H-bonds to the central molecule, without rotating and violating their local repulsive contacts (for partial cages see section 5.2.2).

In the light of the structural characteristics gained from the hydrate and ice crystals (section 5.1), the large spread in the $O \cdots O$ nearest neighbours, 2.5–3.3 Å, indicates that a large number of different local structures are possible. The individual molecules may have H-bond coordinations of up to about six: three and four being the most common. By the nature of thermal fluctuations present, the overall structure is probably of a random character in which almost any local conformation may exist, provided that it complies with the short-range repulsive restrictions (not in the excluded regions). In addition to this, regions of 'free volume' are likely to be present. These are regions in which not all the repulsive limitations are encountered around a water molecule, hence, it is able to move into the unrestricted space. Examples of this are found in hydrate structures (see figure 20(d)).

Between ice-Ih and liquid water there is an increase of approximately 0.1 Å (2.76–2.85 Å) in the average length of the H-bonds. This corresponds to an average decrease in the $H2–O2 \cdots O1$ remote-neighbour angles of about 7–8° (figure 23). The $O \cdots O \cdots O$ angles may also decrease by this amount, which allows the water molecules to pack more closely than in ice-Ih: the volume decrease in going from ice-Ih to liquid water is about 8 per cent.

5.4. *Interactions with Apolar Groups*

Most of the water–apolar $O \cdots CH_x$ contacts in hydrate crystals are seen to be larger than 3.5 Å, the expected van-der-Waals contact ($C = 2.0$ Å for methyls and $O = 1.5$ Å). However, short contacts of around 3.2 Å are sometimes observed, comparable with the short $O \cdots O$ non-bonded distances. The water–apolar contacts can be rationalized in terms of two main interactions: $O(W) \cdots C$ and $O(W) \cdots H$ depending on the relative orientations of the CH_x group and the water molecule. Where the water is situated in a position between the hydrogens of CH_x, as in figure 28(a), then the main

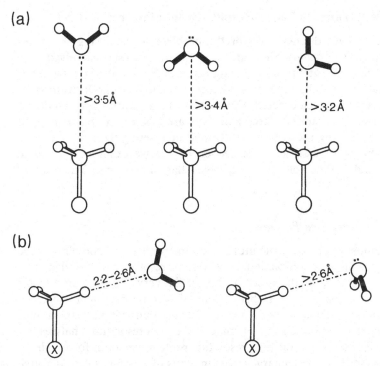

Figure 28. Non-bonded contacts of waters to apolar CH_x groups: (a) away from H; (b) in line with H.

interaction is between the oxygen and carbon. The $O \cdots C$ contact distance also depends on how the water molecule is orientated: when the lone pairs point away from the carbon, distances down to approximately 3.2 Å are observed indicating that a van der Waals carbon radius of 1.7 Å would be more suitable than 2.0 Å. When the lone pairs point towards the carbon, distances of 3.5 Å or greater occur.

When the water molecule is located in a direction that is almost linear with the C–H vector (figure 28(b)), the main interaction is between the hydrogen of the CH_x and the water. Here, an opposite effect relative to the $O \cdots C$ contacts is observed: with the lone pairs pointing away from the hydrogen, the closest $O(W) \cdots H$ contacts are usually between 2.6–2.7 Å. However, when the lone pairs point towards the hydrogen the $O(W) \cdots H$ distances are very often less than 2.6 Å. The values observed appear to depend on the electronegativity of the adjacent group (X) attached to the CH_x. In several hydrates, $O(W) \cdots H$ distances down to 2.2–2.3 Å occur and these contacts are probably weak C-H $\cdots$ O hydrogen bonding interactions [120]. Where the CH_x groups are attached to saturated carbons, $O(W) \cdots H$ contacts of greater than 2.4 Å are commonly observed.

6. Analysis of Solvent in Larger Hydrate Crystals: Proteins and DNA

In this section we discuss some of the problems in analysing the water structure in larger systems. There are basically two main steps involved: first we must locate the water positions, and then secondly, identify possible water structure(s). In small hydrates this procedure is fairly straightforward since the water is usually well ordered. However, for larger hydrates a proportion of the solvent is often disordered and there are inherent problems in both stages: in examining the diffuse density and interpreting the structure from a large number of assigned sites. In section 6.2 we examine both these problems and outline some ways in which they may be tackled in larger hydrates.

6.1. *Current Status and Progress*

A large number of protein structures and several nucleic acid complexes have been solved with X-ray diffraction. However, the resolution achieved in most of these analyses is in the order of 1.5–2.5 Å. Very few macromolecular structures have been studied to better than 1.5 Å resolution, mainly because a large number of the crystals do not diffract very well at higher resolutions. Here, a short summary is given of some of the high resolution analyses.

Table 3 lists several of the high-resolution protein structures for which the solvent regions have been characterized in terms of assigned sites. Although some local water networks have been located in the crevices, over some parts of the surfaces and between the biomolecules, specific details of the water structure (hydrogen positions, H-bond networks, etc.) have not as yet been reported. In crambin crystals, approximately 80 per cent of the solvent appeared to be ordered and a total of 73 solvent sites were assigned including four ethanol molecules. Several pentagonal rings have been reported to occur around some of the surface apolar groups [129]. In the porcine 2Zn insulin analysis, a high percentage of the solvent – approximately 70 per cent – was accounted for with approximately 340 solvent sites, the majority of which were assigned unit occupancies. In the analysis of the solvent in rubredoxin crystals, the individual occupancies of the water sites were refined and a cut-off level (occupancies more than 0.3) was used to distinguish between solvent and background. A smaller percentage of the solvent was modelled – about 50 per cent – using 127 sites. From these examples, it is apparent that currently there is no uniform or consistent method being used for solvent analysis.

At resolutions of 1.5–2.5 Å, only the more ordered solvent sites on the surface of the biomolecule and a few well-ordered sites in the second layer are visible. Very little information about the solvent structure is obtainable at lower resolutions, of more than 2.5 Å, because of problems associated with the weak solvent density observed and resolution limits of the data ($O \cdots O$ H-bonds are about 2.5 to 3.1 Å long).

Table 3. *Examples of protein structures in which the solvent has been characterized*

Crystal hydrate	Radiation (X, X-ray; N, neutron)	Resolution (Å) X	N	No of waters	Refs	Comments
Rubredoxin (54 residues)	X	1.2		127	88	Extensive H-bonding network over surface of protein; 83 in first shell, 40 in second shell
2Zn insulin (51 residues, 2 chains)	X, N	1.5, 1.2, 1.2	2.2	~340	121–124	About 340 waters assigned in one X-ray model [121], and ~280 in another [122].
Crambin (46 residues)	X, N	0.9,	1.4, 1.1†	~80	108, 125	Most of the solvent is ordered; 4 ethanols located in X-ray; several pentagonal rings of waters around hydrophobic groups
Bovine pancreatic trypsin inhibitor (BPTI) (58 residues)	X, N	1.0,	1.8	63	126, 127	Sixty-three sites assigned in joint X-ray/neutron model
Avian polypeptide hormone (APP)X (36 residues)	X	1.0		>90	128	Located mainly on the surface

† A further set of neutron data to 1.1 Å resolution has been collected since this manuscript was prepared.

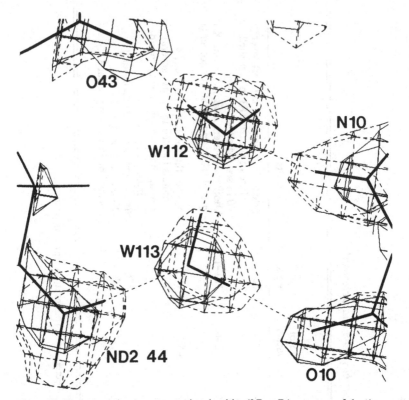

Figure 29. Electron and neutron scattering densities $(2F_O - F_C)$ over two of the 'internal' water molecules, W112 and W113, in the BPTI protein structure. The resolutions of the X-ray and neutron data are 1.0 and 1.8 Å respectively. Full lines: X-ray contours; broken lines: neutron contours.

Neutron studies of larger systems are difficult to perform for several reasons. Firstly, hydrogen possesses a high incoherent scattering cross-section which, in hydrogenated structures, contributes significantly to the background levels, hence the signal-to-noise ratio is poor. This effect may be significantly reduced when the hydrogens are exchanged for deuteriums, which have a relatively low incoherent component and a coherent part comparable with carbon and oxygen. Secondly, at lower resolutions the hydrogens (usually replaced by deuteriums) are not very easy to locate with any confidence, especially if they are disordered. Figure 29 shows an example of two very well-ordered water molecules in a 1.8 Å neutron analysis of bovine pancreatic trypsin inhibitor (BPTI) [127] in which the deuterium positions are just visible. Only the unambiguous water deuteriums attached to non-protonated polar groups (carbonyls, etc.), were really visible in this analysis; the majority remained undetected. A method under current development [130] that may be used to enhance the visibility of the solvent hydrogen positions, is to calculate a difference Fourier synthesis using data collected from deuterated and hydrogenated crystals of the same macromolecule: $F_{O(D)} - F_{O(H)}$ coefficients. This method has been used at 2.1 Å resolution in the solvent analysis of trypsin [131] and most of the hydrogens of the internal and strongly localized surface waters could be identified. Similar techniques are being applied to BPTI and lysozyme (unpublished).

6.2. *Problems Encountered in Solvent Analyses*

Ideally, we would like to examine the solvent structure in large hydrate crystals using atomic-resolution neutron-diffraction data. The best we can achieve at the moment is restricted to medium-sized systems, but even here there are some considerable problems to address in the analysis and interpretation of the solvent density. We now discuss some of these problems.

6.2.1. *Disordered Solvent Density.* Interpreting the more diffuse solvent density associated with disorder is one of the major problems in a solvent analysis. Methods of improving the phases such as density modification [132, 133] may help here. However, there still remains the question – how can these space–time averaged density distributions be modelled? Does one use a system of point sites with a temperature factor smearing function or a more complicated model taking some or all of the dynamics of the system into consideration? The latter model is inherently more complicated, but may possibly be coupled with computer simulation methods such as Monte Carlo or molecular dynamics. Although some progress has been made in this area for smaller systems [35–41], solvent modelling in larger systems [34, 134] appears to be more difficult to tackle (e.g., complexity of potential functions used, length of computational times). Thus, at present, we have to fall back on the former method bearing in mind that we are dealing with a dynamical system. Using this more 'static' approach, the following steps may be used in analysing solvent density:

Stage 1. Assign *main* sites representing the more ordered solvent positions to the regions of well-defined density and include them in least-squares refinement. From these sites the better-defined solvent networks can be formulated (section 6.3).

Stage 2. With the ordered sites located, the diffuse and elongated solvent density around and between the main sites can be examined. These regions can then be modelled by including 'continuous' sites (representing the continuous density – see figure 30) at intervals of about one-third of the data resolution.

Stage 3. For the continuous bulk solvent density a three-dimensional grid of sites is now commonly used [135], with each point having a temperature 'smearing' factor that effectively transforms the grid into a continuum. If necessary, this model can also be applied to regions of diffuse density that are not in the bulk solvent.

The first two steps have been used in the analysis of the coenzyme B_{12}

124   *Hugh Savage*

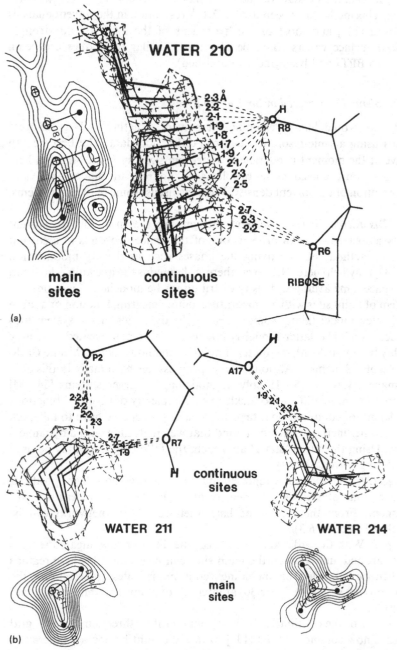

Figure 30. Interpretation of the solvent regions in the coenzyme B$_{12}$ structural analysis: (a) over the 210 solvent region; (b) over the 211 and 214 solvent regions. 'Main' sites initially assigned to the well-defined solvent density. 'Continuous' sites assigned to the elongated and diffuse regions of solvent density.

solvent [96,98] (examples shown in figure 30) and can be readily applied to medium-sized systems. However, when the background noise level of the solvent density increases, as is often the case in protein maps, the analysis in stage 2 is severely hampered and only stage 1 (in conjunction with 3) can be used with any reliability. One way in which this problem may be improved is by using as many different sets of diffraction data (both neutron and X-ray) as possible. For example, four different data sets (one neutron and three X-ray) were used in the coenzyme B_{12} solvent analysis. It was observed in this system that, *when present*, nearly all the alternative solvent networks occupied almost identical positions in each of the four models. However, the individual networks were seen to have *different* partial occupancies among the four *separate* models. Figure 31 shows one example of this, in relation to the disorder of one of the side groups, N40, and its effect on the interpretation of the surrounding solvent. The N40 side chain is disordered between at least two extreme positions (N40 and N640, approximately 1.8 Å apart) and a separate water network is associated with each of the two positions: network A for N40 and network E for N640. In the neutron model the networks have occupancies of 0.6 (A) and 0.4 (E), while in two of the X-ray models the occupancies are 0.9 (A) and 0.1 (E). However, in the third X-ray model the occupancy values are reversed: approximately 0.1 (A) and 0.9 (E). Thus, evidence for the acceptance of weaker peaks as possible solvent sites can be gained from a comparative analysis of different sets of data on the same crystal system.

Similar solvent networks may be present in each of the individual crystals for a given hydrate, but they may well have differing partial occupancies within each crystal. The reason for this may well stem from slight differences in the physical conditions of the crystals used, such as pH, temperature, solvent composition (ions etc.).

6.2.2. *Overlap of Solvent Peaks.* In neutron studies of hydrogenous solvents, there is an additional problem of the negative scattering of hydrogens (associated with a phase change in the scattering process). Overlap of any disordered solvent oxygens (or other positive scatterers) and hydrogens leads to either partial or complete disappearance of solvent peaks. This problem can be partially remedied by replacing the exchangeable hydrogens for positively scattering deuteriums, thus making their detection possible. On the other hand, the combined use of H and D neutron data (assuming the disorder of the solvent, in terms of the site occupancies, is the *same* in the *different* crystals), can be used to improve the visibility of hydrogens (see above).

6.2.3. *Chemical Identity of Solvent Peaks.* This is a problem in relation to formulating solvent networks and deriving solvent structures. A solvent peak in the density map may be a solvent species other than water, for example

(a) neutron

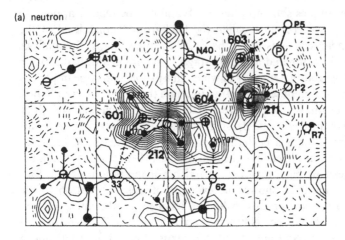

(b) X-ray2

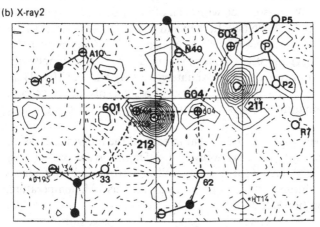

Figure 31. Solvent density around the disordered side chain (N40) of the coenzyme B_{12} molecule: (a) neutron; (b) X-ray2 F_O-F_C difference Fourier maps (the N640 position is not shown but lies approximately 1.8 Å behind N40). Two independently, partially occupied solvent networks are present. Network A: Waters 211 and 212, occupancies–0.9 (X-ray2), 0.6 (neutron) and Network E: Waters 601, 603 and 604, occupancies–0.1 (X-ray2), 0.4 (neutron).

a salt ion used in the crystallization procedure. Three possible ways that can be used to resolve this ambiguity are: (1) Try to formulate water networks over the relevant region and where they are seen to break down, a 'foreign' solvent molecule or possible disorder of the biomolecule must be suspected and taken into consideration. (2) Analyse the peak heights in the suspect region. For solvent species containing heavier atoms, such as phosphate groups, this approach is usually easier for X-rays than for neutrons. Where the solvent density is significantly higher than what would be expected from one water molecule with unit occupancy in a particular region, then the possibility of the inclusion of a non-water species must be considered. (3) Use isotopic exchange in neutron analyses: here certain atomic species can be substituted for an isotope that has different neutron scattering properties. One example is the exchange of the hydrogens for the deuteriums giving the solvent molecules a higher scattering value, thus enhancing their visibility in the maps. This method has also been used in locating the positions of deuterated ethanol molecules (C_2D_5OH) around lysozyme [136].

At fairly low resolutions of approximately 2.0 Å, 'foreign' solvent molecules have been observed in several crystal structures: for example, sulphates in ribonuclease A [137,138], ethanol in lysozyme and α-lytic protease [139]. Some partially occupied 'foreign' sites have been found in several systems of higher resolution, approximately 1.0 Å: for example, phosphate in BPTI [127], ethanol in crambin [108] and acetone in coenzyme B_{12} [96–8].

6.2.4. Disorder of Flexible Side Groups. Disorder of individual atomic groups of the c/molecule usually gives rise to associated disorder of surrounding solvent. This type of disorder adds many more degrees of complexity to the task of formulating local solvent networks (see below), since all the possible cases must be accounted for. In larger structures, it becomes a significant problem. Before an in-depth analysis of the solvent regions can proceed, details of all the possible disordered conformations of the c/molecule (side chains etc.) should be examined.

6.2.5. Occupancy and Temperature Factors. In least-squares refinement, the occupancy and thermal-factor parameters are correlated. Except for very well-ordered solvent peaks it is difficult to obtain reliable values for these parameters in the more disordered solvent regions. Initial occupancies of the solvent sites are very often estimated from their peak heights in the density maps and then included in least-squares refinement. Here, alternate cycles of refinement are usually performed in which one of the two parameters is held fixed while the other is allowed to vary. What often happens in the refinement is that the variable parameter (e.g., temperature factor) adjusts accordingly to the 'fixed' value of the other parameter (occupancy). Thus, if an initial occupancy value is set too low (from the map) for a peak that is known to be very diffuse, then the temperature factor which should have a large value to accommodate this, turns out to be too small (and vice versa). In the larger systems the error estimates (e.s.d.) for the solvent sites are usually large, in the range of 0.2 upwards for the occupancy values, and it is difficult to place very much confidence in their values as assessed from least squares. An alternative method which is sometimes used, is to fix all (or most of) the solvent occupancies at 1.0 and then assess the order/disorder character from the adjustment of the temperature factors only.

Most models assume the solvent density to be either iso- or anisotropically distributed around a given point. This type of model is readily applicable to the well-ordered solvent peaks, but not for disordered regions where elongated and amoeboid shapes of solvent density are known to be present. A better estimate of the occupancies might be obtained by calculations of the electron or neutron densities of the solvent peaks and their correlation with estimates of the temperature factors from the volumes or 'spread' of the peaks in the maps.

6.3. *Formulation of Water Network Structures*

There are several stereochemical constraints and restraints that can be applied in the formulation of individual water networks. The primary constraints are the basic covalent geometries of the water molecules, which can be given the average values found in small molecule surveys: O-H bond lengths of 0.96 Å and H-O-H angles of 107° [46]. The restraints that can be considered are those of H-bonding and repulsions (section 5). They are quite flexible, but at the same time have 'limits' (repulsive) beyond which no structural geometry is acceptable. Restraints used in the formulation of the intermolecular structure can be divided into two areas:

(*A*) *Geometrical Structural Restraints*

(1) $O \cdots O$ distances for H-bonding lie between 2.5 and 3.2 Å. Distances less than 2.5 Å considered as mutually exclusive sites and alternative networks formulated.

(2–5) Short-range contacts around each water molecule must observe the four repulsive restraints criteria, RR1, RR2, RR3 and RR4, given in section 5.1.2.

The $O\text{-}H\cdots O$ H-bond angles should be optimized to be as close to 180° as possible. Where local repulsive contacts prevent this, values within the accepted limits of H-bond bending are allowed.

(*B*) *Criteria Related to Crystal Factors*

(6) Occupancies: where possible, networks formulated from solvent sites with similar values.

(7) Disorder of the c/molecule, in particular the side chains. Surrounding solvent networks formulated for each of the disordered positions.

(8) Presence of a foreign solvent molecule (sulphate, metal ion, etc.) that may have been used in the crystallization procedure: continual checks must be made to see if some of the assigned solvent sites might correspond to a solvent species other than water.

In several cases the position of the hydrogen may have to be inferred due to the solvent being (1) uninterpretable and/or (2) disordered. They should be incorporated such that they do not violate any of the above criteria, though clearly this may fail to result in unambiguous locations.

6.4. *Strategy and Requirements for a Solvent Analysis*

Essentially this applies to medium and larger hydrate crystals. Primarily, high-resolution data – better than 1.0 Å resolution – should be obtained, both X-rays and neutrons being used to locate the oxygen and the hydrogen positions of the individual water molecules in a complementary manner. To reduce the thermal disorder in the crystals, it would be advantageous to collect the data at a low temperature, 0 °C or below, providing the crystals remain stable and continue to diffract. In addition, several sets of data (neutron and X-ray) on the same system should be collected and if hydrogenated and deuterated neutron data are also available, then these data can be combined to locate the hydrogen positions more precisely.

In the analysis and modelling of the solvent density itself, an integrated approach may be used in which all the methods outlined in the previous section are applied in an iterative manner with respect to the differing levels of the complexity in the solvent disorder. In the first instant, the main solvent networks around the c/molecule – essentially the first layer of hydration – can be derived. Where these networks are seen to break down, for example due to side chain disorder and inclusion of foreign species, the solvent density over these regions may be re-examined and further networks constructed. This process may then be repeated until all the observed density is accounted for using interpretations that fall within the acceptable limits of geometrical 'restraints' obtained from the smaller systems.

6.5. Use of X-ray Diffraction Data

In larger hydrate systems, it may be easier (at least initially) to analyse the solvent using X-ray diffraction. Many more structures are examined by X-rays than by neutrons. In addition to this, the solvent regions in large neutron structures are more difficult to analyse. One major problem is the chemical identity of the density itself: for example, is a positive peak an O or a D? In most X-ray hydrate structures, this problem is more straightforward since only the heavier oxygen atoms (and possibly other heavier species) are located with any reliability (section 3.3): *Given a set of solvent oxygen positions from an X-ray analysis*, we may use the criteria outlined in section 6.3 as restraints along with the addition of hydrogen atoms to the experimental oxygen sites, in the interpretation of water structure(s). Several examples of this approach in medium-sized systems and one in a protein hydrate are now presented.

The crystal hydrate of deoxydinucleotide phosphate d(CpG) proflavine complex contains about 27 water molecules [100, 101]. Using $O\cdots O$ distance criteria of less than 3.3 Å to represent H-bonds, a number of these water molecules appear to form four H-bonded pentagonal rings (I–IV) in one of the solvent regions (see figure 32(a)). A fifth pentagonal ring (V) which includes a phosphate oxygen, is also present. Keeping the experimental oxygen positions fixed, hydrogen positions can be assigned that do not violate the repulsive restraints criteria. One of several possible networks is shown in figure 32(b). However, in order to do this, several of the proposed H-bonds ($O\cdots O$ less than 3.3 Å) cannot be considered as such, otherwise

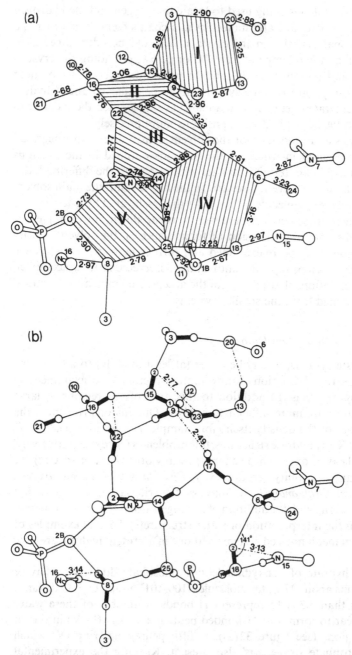

Figure 32. Interpretation of water structure in the crystal structure of d(CpG) proflavine complex. (a) X-ray structure: solvent oxygen positions (and a phosphate oxygen) form five pentagonal rings (shaded). (b) Possible water structure derived from placing hydrogens around the experimental oxygen sites in compliance with the four repulsive restraints. Distances in ångströms.

some very close H2···O1 and H···H contacts occur. Consequently, some of the pentagonal rings are probably *not* five-membered *H-bonded* rings. In ring IV, molecules W18 and W25 form a triangular conformation with phosphate oxygen O1B and the three O···O distances are 2.67, 2.92 and 3.23 Å (figure 32(a)). Assuming the two shorter distances are H-bonds (to O1B), it is impossible to place a hydrogen between W18 and W25 without violating local H2···O1 and H···H minimum contacts (angle O1B···W25···W18 = 51°). The W18···W25 distance is relatively long and when assigned as a close non-bonded O···O contact, none of the local repulsive contacts around molecules W18 and W25 are violated. Thus in terms of H-bonding, ring IV appears to be a six-membered ring incorporating oxygen O1B.

From the given X-ray sites, it appears to be doubtful whether rings I, II and III form as such (figure 32(a)). The H-bond W9···W17 is very long (H···O ≃ 2.5 Å) and bent, while the H-bond W15···W23 appears to be too short (O···O = 2.42 Å) for a normal asymmetric bond. Some of the H2···O1 and H···H contacts around the latter bond are below the allowed minimum values (e.g. H2(W15)···O1(W23) = 2.77 Å). The temperature factors (measure of spread of electron density) of molecules W9, W13, W15, W20 and W23 in rings I, II and III are substantially higher (B = 18–32 Å²) than for the other water sites (B = 10–16 Å²). Alternative partially occupied sites may be present around these molecules and some of these positions may be acceptable in terms of satisfying the repulsive restraints. Whether or not they form five-membered rings, depends on the details of the local short-range interactions. Further analysis of the solvent density is required (along the lines given in section 6.2), before the structural configurations over rings I, II and III can be better assessed.

It is possible to see why some of the H-bonds are relatively long, 2.9–3.2 Å. These are mainly due to restrictions of the H2···O1, N contacts that are close to their minimum values. For example, H2 of water W18 has to rotate away from N15 to relieve the close contact of approximately 2.8 Å between these atoms (when H2 is placed on the W18···W6 line). In figure 32(b), water W18 is orientated such that H1···N15 is greater than 3.1 Å, making the H-bond angle W18–H2···W6 substantially less than 180°, resulting in the necessity of a longer O···O distance. Similarly, the H-bond of W8 to phosphate oxygen O2B has to be long and bent in order to relieve the close contact between H1 of water W8 and N16.

About half of the highly occupied water positions located in the original X-ray structural determination of vitamin B_{12} coenzyme [107], have only low occupancies in the neutron structural analysis of this hydrate [96–99]. The remainder were unobserved in the neutron solvent maps. These regions in the X-ray water structure (with unknown hydrogen positions) could be interpreted using the restraints criteria. Figure 33(a) shows the highly occupied X-ray positions in the pocket region (with local distances and angles), while figure

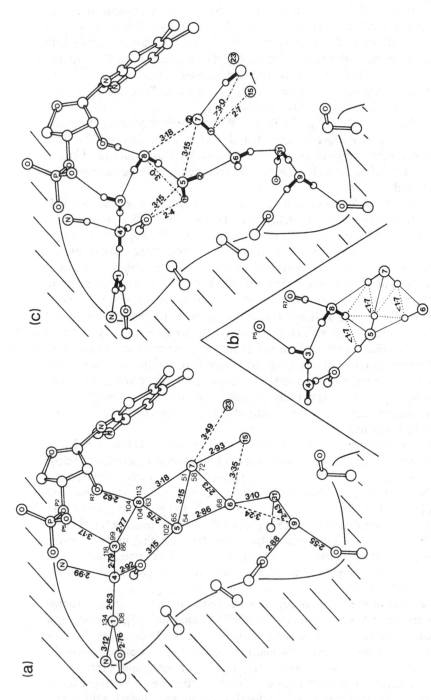

Figure 33. Interpretation of water structure in the original X-ray structure determination of coenzyme B_{12} at 1.2 Å resolution [107]. (a) Distances and angles around the experimental water oxygen sites. (b) Inclusion of hydrogens between $O \cdots O$ distances less than 3.2 Å gives rise to some very short $H \cdots H$ contacts of < 1.7 Å (shown by dotted lines). (c) Possible water structure that complies with local repulsive restraints. Distances

33(c) shows one of several possible water structures that complies with the repulsive limitations. One region of particular interest is that over waters W5, W6, W7 and W8. When H-bonds are assigned using only the criterion of $O \cdots O$ distances between 2.5 and 3.3 Å, two three-membered rings with $O \cdots O \cdots O$ angles from 51–68° are present (figure 33(a)). However, in trying to place hydrogens between the five 'assigned $O \cdots O$ H-bonds' of these two rings (W5, W7, W8 and W5, W6, W7), some very short $H2 \cdots O1$ (less than 2.8 Å) and $H \cdots H$ contacts (less than 1.7 Å) are unavoidable (see figure 33(b)). When some of the $O \cdots O$ distances greater than 3.1 Å are considered as non-bonded contacts, hydrogens may be placed such that the local $H2 \cdots O1$ and $H \cdots H$ contacts are all satisfied. An H-bonded structure is possible, in which several of the water hydrogens do not participate in H-bonds. This phenomenon is seen in the experimental neutron analysis of this system (figure 11). The very weak H-bond between W5 and carbonyl oxygen O62 of 3.15 Å appears to occur because of the minimum $H2 \cdots O1$ contact between waters W5 and W8 (see figure 33(c)). The solvent density between waters W15 and W23 is continuous and to maintain minimum $H2 \cdots O1$ contacts around a water in this region H-bondng to W7, its position must lie closer to W23 than W15 (figure 33(c)).

The structure of [Phe⁴Val⁶] antamanide [102] comprises a cyclic ring of ten amino acid residues. The hydrate crystal contains 12 water molecules per peptide molecule. Six of these are situated above the peptide ring as shown in figure 34(a). At first sight it would appear that water W1, which is located on a two-fold symmetry axis, forms four H-bonds in a tetrahedral arrangement to two nitrogens and waters W2 and W3. However, when the surrounding $O \cdots O$, N distances are examined (figure 34(a)), five distances are shorter than the long H-bonds of 3.16 Å (W1 $\cdots$ W2, W3). In addition to this, waters W2 and W3 (symmetrically equivalent) have four $O \cdots O$, N distances of less than 3.2 Å that are all on the same side of a plane through these water molecules. Clearly not all four of these contacts can be H-bonds, otherwise the inclusion of the hydrogens would lead to exceptionally short $H \cdots H$ contacts down to about 1.5 Å, even when the H-bonds are made as bent as possible (lie on H-bond bending limits curve, figure 14). An interpretation of the structure in this region which satisfies all four of the repulsive restraints can be found if some of the $O \cdots O$ distances greater than 3.1 Å are considered as non-bonding. Figure 34(b) shows a possible structure in which water W2 is removed to relieve short non-bonded $H \cdots H$ contacts that would consequently occur between waters W1 and W2. In this structure there is still one short non-bonded $O \cdots O$ contact of 3.01 Å between W1 and a carbonyl oxygen. To alleviate this contact, it is possible to move W1 by about 0.1 Å (within the mean temperature-factor displacement for this site, approximately 0.3 Å) to the left (A in figure 34(b)).

In doing these structural analyses, details of the disorder of each water site in terms of its temperature factor and surrounding solvent density (elongated

(a)

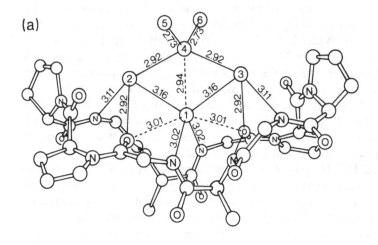

(b)

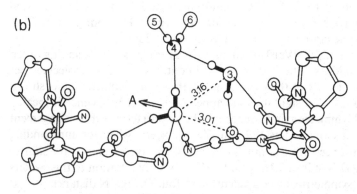

Figure 34. Interpretation of water structure in the crystal structure of [Phe⁴Val⁶] antamanide hydrate. (a) X-ray structural positions of six waters over the cyclic peptide ring. O⋯O, N distances around waters are shown. (c) Possible water structure that complies with the repulsive restraints. To relieve the O⋯O non-bonded contact of 3.01 Å, water 1 may be moved to the left (direction of A). Distances are given in ångströms.

peaks, etc.) should be noted. Where larger temperature factors are present, alternative water sites may be present, from which networks can be formulated that satisfy local repulsive limitations.

Of the several high-resolution protein structures, in which the solvent has been examined (see table 3), porcine 2Zn insulin is one of the most extensively studied systems. At least four different groups have collected diffraction data (X-ray and neutron), some of which extend to as high as 1.1 Å resolution. The solvent regions within the 2Zn insulin crystals have been well characterized in the X-ray analyses. In the 1.5 Å resolution X-ray analysis [140], about 340 solvent sites (all oxygens, apart from 2–3 possible ion sites) were assigned and carefully refined without non-bonded restraints between the

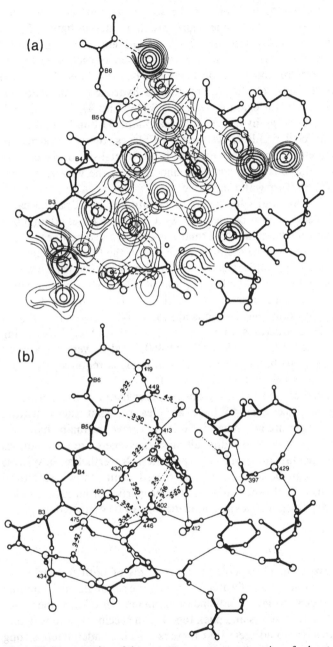

Figure 35. Interpretation of the water structure over one region of solvent density adjacent to residues B3–B7 (molecule 1) in porcine 2Zn insulin crystals. (a) Solvent electron density at 1.5 Å resolution from X-ray refinement studies of this crystal structure [140]. The refined water positions are included and the broken lines represent O···O, N distances of less than 3.4 A. (b) One of several possible water networks constructed in which the repulsive restraints are satisfied. The broken lines in (b) represent non-bonded O···O contacts.

sites. The electron density over one of the solvent regions and an initial structural interpretation of this same region are illustrated in figure 35.

Several alternative sites of less than 2.5 Å are present. A possible water network structure was derived from the well-defined oxygen sites (figure 35(b)), but others (mainly local disorder) may also be formulated. Within the proposed network, some of the waters possess hydrogens which do not participate in H-bonds, for example waters 413, 460 and 475. It is difficult to move these non-bonding hydrogens into reasonable H-bonding positions without violating local repulsive restraints. In places, the water structure appears to be somewhat open, probably due to the actual size of the cavities occupied. That is, there is extra space around some of the waters, but not enough to include another water molecule.

Water molecules 402 and 446 appear to be situated in a 'partial cage' arrangement with several non-bonded $O \cdots O$, C contacts of less than 3.6 Å around them. Several triangular conformations of waters (sometimes including polar protein atoms) are present in which one of the $O \cdots O$ distances is longer than 3.1 Å: for example, waters 402, 412 and 432. The presence of a longer $O \cdots O$ distance allows the formulation of a local structure in which the minimum short-range contacts are satisfied. These apparent three-membered rings with only two H-bonds appear to be a relatively common feature in the hydrate structures studied so far. In the central region involving water molecules 402, 412, 430, 446, 459 and 460, local movements around individual sites seem to be possible (further analysis is required), that are similar to those in coenzyme B_{12} (figure 20(d)).

In order to interpret the solvent regions of larger hydrates in a consistent manner, it appears necessary to have an idea of what the structural characteristics of water are beforehand. To try to derive the main characteristics from such large systems is a difficult task, since there are uncertainties in both the experimental data and its interpretation. Nevertheless, in systems where the solvent can be adequately interpreted using known structural restraints, some useful information about the properties of large assemblies of water molecules (such as packings and *PCFs*) can be extracted.

7. Discussion

Although we have at hand experimental structures of both regularly and irregularly ordered assemblies of water molecules, the rationale of H-bonding *alone* does not appear to give a satisfactory explanation of how the water molecules orientate themselves and pack together in specific positions within the said structures. It would seem that there is a lack of adequately strong stereochemical constraints (which are normally present in covalently bonded structures). However, the above analysis of small and medium neutron hydrate structures has revealed that some can be formulated with respect to short-range non-bonded contacts between the water molecules. These short-range restraints give us a more reasonable picture of the intermolecular structure and they may also be used to assist the interpretation of solvent in larger structures.

The use of O···O distance criteria alone for H-bonding, especially around 3.1–3.5 Å, may be misleading. This is because some of the hydrogen-involved repulsive constraints may be violated for assigned bonds. As with the simpler systems [28–33], the repulsive contacts appear to play a dominant role in determining the final orientational geometries of the local structure. Important contributions appear to come from the asymmetric repulsive character of both the oxygens and the hydrogens. Repulsive forces vary much more rapidly than the longer range attractive interactions and consequently, the ranges of repulsive contact distances observed are more limited than those for the attractive distances.

7.1. *Summary of Short-range Contacts*

The results of the above analyses in sections 4 and 5, indicate that there are at least four different types of repulsive interactions between the oxygens and hydrogens of the water molecules. These are:

RR1, oxygen–oxygen repulsion in the H-bonds;
RR2, oxygen–oxygen next-nearest-neighbour repulsions;
RR3, hydrogen–oxygen (not of H-bond) remote-neighbour repulsions;
RR4, hydrogen–hydrogen repulsion.

The first interaction RR1 tends to force the H-bond to be close to linearity at shorter O···O H-bonding distances (O–H···O angles near to 180°). The second, RR2, appears to be somewhat anisotropic with possible O···O contacts ranging from 3.1 to 3.6 Å: the longer contacts occur in the proximity of the oxygen lone pairs. The minimum H2···O1 remote neighbour contacts, RR3, are in the order of 3.0 Å (O–H = 0.8 Å), which appear to be outside the presently accepted H···O van-der-Waals contacts radii. Minimum H···H contacts of between 2.1 and 2.7 Å are observed for RR4 depending on the mutual orientations of the O–H groups: H···H contacts tend to be shorter for the O–H vectors in the 'head-on' arrangement.

The third and fourth interactions, RR3 and RR4, may be rationalized in terms of assigning an angular-dependent variable van der Waals radius for the water hydrogen. Values in the order of 1.5 Å in the H–O direction (i.e. O–H···H angles less than 60°) and below 1.0 Å in the O–H direction (O–H···H angles greater than 140°), may be appropriate (figure 25(a)). This approach is consistent with the known asphericity of the electron cloud over the polar hydrogen atoms: the electron distribution of the hydrogen is shifted towards the oxygen centre.

A reasonable model for the non-bonded intermolecular contacts will have to take into account the anisotropic effects of both the oxygens and hydrogens. Outside of quantum mechanics, most present-day models treat the short-range contacts as interactions between spherical centres and rely on the more slowly varying attractive interactions (charges, dipoles, etc.) to determine local structure. The data presented in the above sections suggest that this approach is probably an over-simplification of the real situation. The exact physical nature of the RR3 and RR4 contacts are at present not clearly understood and here quantum-mechanical calculations may provide some initial answers.

Within a water structure, *all* four of the repulsive interactions must be satisfied in terms of maintaining the minimum-allowed contacts. Of the four interactions, the $H2\cdots O1$ (RR3) contacts appear to play a prominant role in determining the immediate local orientational geometries. This is particularly the case for the H-bond geometries, which tend to bend to within the limits governed by the $O\cdots O$ repulsion of the H-bond (*H-bond bending limit curve*: figure 14), to accommodate the minimum allowed $H2\cdots O1$ contacts. Examples of these coupled interactions are shown in figures 19 (H-bond geometries) and 20 ($H2\cdots O1$ contacts). The RR2 and RR4 repulsive interactions also play a significant role, particularly when the H-bonds are severely distorted or when there are unavoidable close repulsive contacts, such as in structures at high pressures or where local H-bonding may be weak or non-existent.

7.2. *Computer Simulations: Structural Checks and Restraints*

The repulsive terms used in most of the current water potentials are of the isotropic type: inverse ninth or twelth powers or exponential functions. These are usually placed on the oxygen centres and in a few cases, on the hydrogens as well, for example the MCY potential of Matsuoka *et al.* [73] and EMPWI of Vovelle *et al.* [141]. From the above analysis, it seems appropriate to include asymmetrical repulsive cores for *both* the oxygens and the hydrogens. In the case of the latter, there may be some initial difficulties until the $H2\cdots O1$ type interactions are more clearly understood.

Morse and Rice [142] have tested the accuracy with which some of the effective pair potentials can reproduce several of the experimental high-pressure ice structures: II, VIII and IX. In this analysis three potentials were used.

ST2: point charges, dispersion and inverse repulsion on O;

MCY: point charges, dispersion and exponential repulsion on O and H (analytical fit of parameters to quantum-mechanical calculations of energy surface); and

RSL2: central force model, point charges, dispersion and inverse repulsions for O and H.

The MCY potential was found to give the best agreement for the $O\cdots O\cdots O$ angles, but poor agreement for $O\cdots O$ H-bond distances (volume increases of approximately 20 per cent). The results obtained from the other two were reversed: good agreement on distances, bad on angles. The central-force model predicts the formation of bifurcated H-bonds in ice-VIII which are not found in the experimental structure (figure 5(b)). The inclusion of repulsive centres for the hydrogens in the rigid-body MCY model appears to improve the orientational structure. However, this model is apparently too repulsive (volume increases); probably because the repulsive cores are isotropic and this may be a particular problem for hydrogen which appears to be more anisotropic than the oxygen (see above).

The four repulsive restrictions may be used as limiting restraints in a computer simulation of water structure (Monte Carlo or molecular dynamics [24]), and/or in analysing the results of simulations. Only the water–water geometries that do not violate the minimum-allowed atom–atom contacts would be permitted. An example of the usefulness of restraints can be seen in the examination of the simulated water structure (Monte Carlo) in the amino acid hydrate, L. arginine dihydrate [35] for which an accurate neutron structure is known [113]. Within this hydrate, there are two water molecules per asymmetric unit and these were the only entities allowed to move in the simulation. The polarizable electropole potential of Barnes *et al.* [79, 143] was used to model the water molecules (no repulsive hydrogen centres were included), while point charges and Lennard–Jones functions were used for the amino acid atoms. The resultant simulated water model did not entirely correspond to the neutron structure. The water molecules form a continuous chain of water–water H-bonds down one axis of the crystal (figure 36(a)). In the neutron structure, the hydrogens of this chain all point in one direction, whilst in the simulated model, they were found to point in the opposite direction. The hydrogens do not appear to be disordered in the neutron structure.

An explanation of this apparent difference can be found in an examination of the local $H2 \cdots O1$ and $H \cdots H$ non-bonded contacts of the H-bonded water networks. The contacts in the alternative networks are shown in figure 36(b) for the neutron structure and figure 36(c) for the same structure when the hydrogen positions of the H-bond chain are reversed. In the experimental structure, all the minimum contacts are satisfied, but when the water–water hydrogens are reversed some of the minimum $H2 \cdots O1$ and $H \cdots H$ minimum contacts are substantially violated (figure 35(c)). Thus, the water molecules prefer to adopt only one orientation in this crystal. In the simulation, however, repulsive centres were not included for the hydrogens. Hence, restraints involving van der Waals contacts around the hydrogens (RR3 and RR4) were not operative in the modelled structure and close $H2 \cdots O1$ and $H \cdots H$ contacts were possible, allowing the alternative H-bonding network.

In most simulations (using various potential functions), the level of agreement of the water oxygen positions between experiment and model varies over a large range (0.2–2.0 Å). Agreement for the hydrogens is usually poorer than for the oxygens, with waters often being orientated in different

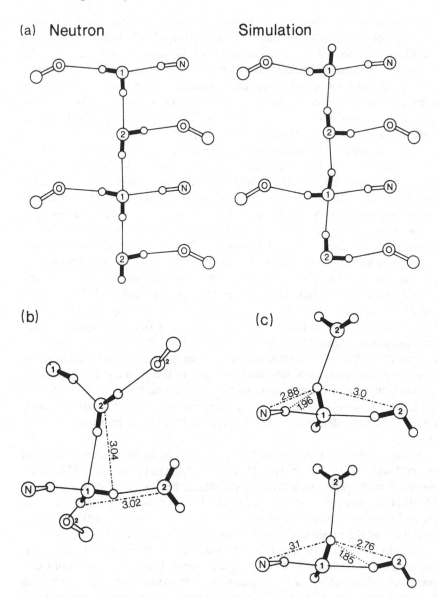

Figure 36. H-bonding, H2···O1 and H···H contacts in L.arginine dihydrate. (a) Schematic diagram of the H-bond networks of waters in the neutron (left) and simulated (right) structures. (b) H2···O1 and H···H contacts around waters in the neutron structure. (c) H2···O1 and H···H contacts for the same structure, but with the chain of water–water H-bonds reversed. The remote H2···O1 contacts are partially satisfied: first by setting H1···O2 (water 2) to 3.0 Å and then setting H1···N (nitrogen) to 3.1 Å. Distances in ångströms.

directions from those of the experimental structure, making alternative H-bonds to other groups. Inclusion of asymmetric repulsive cores for both the oxygens and hydrogens in potential functions or the use of the short-range contacts as limiting restraints may improve agreement.

7.3. *General Comments*

When details of the short-range repulsive forces are considered, in addition to H-bonding, there emerges a reasonably consistent picture of the organization of water structure. The geometrical evidence discussed in section 4 suggests that many of the observed structural features in various water phases can be explained qualitatively assuming the repulsive cores of the oxygen and hydrogen atoms are asymmetrical. The resulting repulsive restraints (section 7.1) appear to largely control the orientational structure. For instance, the presence of the basic tetrahedral characteristics can be related to an optimization of the repulsive interactions (particularly the remote $H2 \cdots O1$ contacts) and a maximization of the number (four) and strengths (2.7–2.8 Å) of H-bonds formed. Shorter H-bonds tend to give rise to trigonally coordinated waters while long bonds allow significant distortions from tetrahedrality. The remote $H2 \cdots O1$ minimum contacts also place restrictions on the smallest ring structures that can be formed, in which normal H-bonds are formed (2.7–2.8 Å): five-membered rings are the smallest possible for unstrained contacts. The occurrence of relatively 'stable' bifurcated H-bonds seems to result from the three-centred hydrogens being held in place by local $H2 \cdots O1$ and $H \cdots H$ repulsions.

In the ice polymorphs, the remote $H \cdots O$ and $O \cdots O$ repulsive contacts become fully strained in the various phases, as higher pressures are encountered. Decreases in volume are achieved by adopting structures that have large deviations from tetrahedrality and/or form interdigitating H-bond lattices (self-clathrates). Partial clathrate cages in which individual water molecules are surrounded by several non-bonded water molecules, are seen to be present in hydrates at ambient pressures. Details concerning the dynamics of water molecules from diffraction measurements are limited, since only a space–time averaged picture of the system is given. Nevertheless, some suggestions are apparent in solvent regions where alternative (disordered) positions are assigned around well-defined water sites. Local repulsive restrictions are not usually present in the direction of possible movement. Thus, under thermal fluctuations, a water may move around within the unrestricted 'free' volume into other positions. Similar movements probably occur in the liquid.

Inclusion of the short-range repulsive contacts provides us with a better understanding of the underlying packing features within known water structures. From these considerations, it appears necessary to revise our current opinions and update some of the current potential functions used in simulations of water and aqueous systems. It may also be necessary to place more emphasis on the short-range intermolecular contacts in other hydrogen bonded structures, for example in biological structures such as proteins and DNA.

Acknowledgments

I would like to thank John Finney for his comments and discussions during the preparation of this manuscript. I also express my thanks to Jeremy Baum, David R. Davies, Guy Dodson, Jenny Glusker, Dorothy Hodgkin, Ted Prince, Walt Stevens, Alex Wlodawer and Joe Zaccai for their continued interest, useful discussions and helpful advice.

Appendix

Listed below is a bibliography of the crystal hydrate structures examined in this article. Included in brackets are the number of unique water molecules contained in each hydrate.

A1. *Neutron Structures*

Ice-Ih (1): references 54 and 55.
Ice-II (2): reference 58.
Ice-IX(III) (2): reference 60.
Ice-V (4): reference 63.
Ice-VI (2): reference 53.
Ice-VII (1): references 53 and 67.
Ice-VIII (1): references 53 and 65.
L-arginine . 2H$_2$O (2): reference 113.
L-lysine monohydrochloride dihydrate (2): *Acta Crystallogr.* **B28**, 3207–14 (1972).
L-serine monohydrate (1): *Acta Crystallogr.* **B29**, 876–84 (1973).
L-asparagine monohydrate (1): *Acta Crystallogr.* **B28**, 3006–13 (1972).
L-histidine hydrochloride monohydrate (1): *Acta Crystallogr.* **B33**, 654–9 (1977).
L-cysteic acid monohydrate (1): *Acta Crystallogr.* **B29**, 1167–70 (1973).
Aqua (L-glutamato) cadmium(II) hydrate (2): *Acta Crystallogr.* **B33**, 801–9 (1977).
Dipeptide glycylglycine monohydrochloride monohydrate (1): *Acta Crystallogr.* **B28**, 2083–90 (1972).
Bis-(-L-histidinato)cadmium dihydrate (2): *Acta Crystallogr.* **B32**, 2803–6 (1976).
α-oxalic acid dihyrate (1): *Acta Crystallogr.* **B25**, 2437–51 (1969).
β-oxalic acid dihyrate (1): *Acta Crystallogr.* **B25**, 2442–51 (1969).

Dialuric acid monohydrate (1): *Acta Crystallogr.* B25, 1970–78 (1969).

Violuric acid monohydrate (1): *Acta Crystallogr.* 17, 415–420 (1964).

Ammonium oxalate monohydrate (1): *Acta Crystallogr.* B28, 3340–51 (1972).

β-maltose monohydrate (1): *Acta Crystallogr.* B33, 2490–5 (1977).

Aqua(L-glutamato)cadmium(II) hydrate (1): *Acta Crystallogr.* B33, 801–9 (1977).

α-L-rhamnose monohydrate (1): *Acta Crystallogr.* B34, 2551–5 (1978).

Trisodium tris (oxydiacetato)cerate(III) nonahydrate (9): *Acta Crystallogr.* B32, 3066–77 (1976).

Disodium dihydrogen-silicate tetrahydrate (4): *Acta Crystallogr.* C41, 638–41 (1985).

$Na[Sm(C_{10}H_{12}O_8)(H_2O)_3] \cdot 5H_2O(8)$: *Acta Crystallogr.* C40, 1687–93 (1984).

Sodium hydrogen oxalate monohydrate (1): *Acta Crystallogr.* C40, 1800–1803 (1984).

$Ca(H_2PO_4) \cdot H_2O$ (1): *Acta Crystallogr.* B31, 9–12 (1975).

$La_2Mg_3(NO_3)_{12} \cdot 24H_2O$ (4): *Acta Crystallogr.* B33, 3933–6 (1977).

$MgSO_4 \cdot 4H_2O$ (4): *Acta Crystallogr.* 17, 863–9 (1964).

$MgSO_3 \cdot 6H_2O$ (2): *Acta Crystallogr.* C40, 584–6 (1984).

$MgS_2O_3 \cdot 6H_2O$ (3): *Acta Crystallogr.* C39, 515–18 (1983).

$LiNO_3 \cdot 3H_2O$ (2): *Acta Crystallogr.* B36, 1032–40 (1980).

$Al(NO_3)_3 \cdot 9H_2O$ (9): *Acta Crystallogr.* C39, 925–30 (1983).

Barbituric acid dihydrate (2): *Acta Crystallogr.* B33, 1655–60 (1977).

5-Nitro-1-(β-D-ribosyluronic acid)-uracil monohydrate (1): *Acta Crystallogr.* B35, 1388–94 (1979).

α-cyclodextrin (6): reference 93.

β-cyclodextrin (11–12): reference 95.

Vitamin B_{12} coenzyme (14–18): references 96–9.

A2. *X-ray Structures*

Ice-IV (2): reference 61.

DL-homoproline tetrahydrate (4): *Acta Crystallogr.* B35, 396–8 (1979).

Drug-deoxydinucleoside phosphate d(CpG) complex (27): reference 100.

Vitamin B_{12} coenzyme (16–17): references 96, 97, 107.

β-cyclodextrin (12): reference 94.

Porcine 2Zn insulin ($\sim$ 280): Coordinates from G. G. Dodson, Chemistry Department, University of York, York YO1 5DD, UK.

References

1. J. D. Bernal & R. H. Fowler, *J. Chem. Phys.* 1, 515–48 (1933).
2. *Physics of Ice* (N. Riehl, B. Bullemer & H. Engelhardt, eds), Plenum Press, New York, 1969.

144   *Hugh Savage*

3. D. Eisenberg & W. Kauzmann, *Structure and Properties of Water*, Oxford University Press, London, 1969.
4. *Water: A Comprehensive Treatise*, Vols 1–7 (F. Franks, ed.), Plenum Press, New York, 1972–82.
5. *Water and Aqueous Solutions: Structure, Thermodynamics and Transport Processes*, vol. VIII (R. A. Horne, ed.), Wiley, New York, 1972.
6. *Physics and Chemistry of Ice* (E. Whalley, S. J. Jones & L. W. Gold, eds.), Royal Society of Canada, Ottawa, 1973.
7. I. D. Kuntz & W. Kauzmann, *Adv. Protein Chem.* **28**, 239–345 (1973).
8. P. V. Hobbs, *Ice Physics*, Clarendon Press, Oxford, 1974.
9. A. Ben-Naim, *Water and Aqueous Solutions: Introduction to Molecular Theory*, vol. XIV, Plenum Press, New York, 1974.
10. *Structure of Water and Aqueous Solutions* (W. A. P. Luck, ed.). Verlag Chemie-Physik, Weinheim, W. Germany, 1974.
11. R. Cooke & I. D. Kuntz, *Annu. Rev. Biophys. Bioeng.* **3**, 95–126 (1974).
12. J. L. Finney, *Philos. Trans. R. Soc. London* B **278**, 3–32 (1977).
13. J. T. Edsall & H. A. McKenzie, *Adv. Biophys.* **10**, 137–207 (1978).
14. J. L. Finney, J. M. Goodfellow & P. L. Poole, in *Structural Molecular Biology* (D. B. Davies, W. S. Saenger & S. S. Danyluk, eds.), Plenum Press, New York, 1982.
15. J. T. Edsall & H. A. McKenzie, *Adv. Biophys.* **16**, 53–183 (1983).
16. B. Kamb, in *Crystallography in North America* (D. McLahlan and J. P. Glusker (eds.) American Crystallographic Association, 1983, pp. 336–42.
17. Workshop on Water, *J. Physique* **45**, Suppl. C7 (1984).
18. *Water Science Reviews* 1, (1985).
19. H. S. Frank, in *Water: A Comprehensive Treatise*, vol. 1 (F. Franks, ed.), Plenum Press, New York, 1972, p. 515.
20. F. H. Stillinger, *Adv. Chem. Phys.* **31**, 1–101 (1975).
21. G. Nemethy, W. J. Peer & H. A. Sheraga, *Annu. Rev. Biophys. Bioeng.* **10**, 459–97 (1981).
22. J. R. Reimers, R. O. Watts & M. L. Klein, *Chem. Phys.* **64**, 95–114 (1982).
23. D. L. Beveridge, M. Mezei, P. K. Methrotha, F. Marchese, G. Ravishankar, T. Vasu & S. Swaminathan, in *Molecular Based Study and Prediction of Fluid Properties, Advances in Chemistry Series*, (J. M. Haile & G. A. Monsoori, eds), American Chemical Society, Washington DC, 1983.
24. D. W. Wood, in *Water: A Comprehensive Treatise*, Vol. 6 (F. Franks, ed.), Plenum Press, New York, 1979, p. 279.
25. J. L. Finney, J. E. Quinn & J. O. Baum, *Water Science Reviews* 1, 93–170 (1985).
26. B. W. Matthews, *J. Mol. Biol.* **33**, 491–7 (1968).
27. L. Pauling, *The Nature of the Chemical Bond*, Cornell University Press, Ithaca, New York, 1960.
28. M. P. Tosi & F. G. Fumi, *J. Phys. Chem. Solids* **23**, 359–366 (1962).
29. J. D. Bernal, *Proc. R. Soc., London* A **280**, 299–322 (1964).
30. D. Chandler, *Acc. Chem. Res.* **7**, 246–251 (1974).
31. J. P. Hansen & I. R. McDonald, *Theory of Simple Liquids*, Academic Press, New York, 1976.
32. R. Narayan & S. Ramaseshan, *Phys. Rev. Lett.* **42**, 922–966 (1979).
33. D. Chandler, J. D. Weeks & H. C. Andersen, *Science* **220**, 787–794 (1983).

34. A. T. Hagler & J. Moult, *Nature* **272**, 222–6 (1978).
35. J. M. Goodfellow, J. L. Finney & P. Barnes, *Proc. R. Soc., London* B **214**, 213–28 (1982).
36. K. S. Kim, G. Gorongiu & E. Clementi, *J. Biomol. Struct. Dyn.* **1**, 263–85 (1983).
37. M. Mezei, D. L. Beveridge, H. M. Berman, J. M. Goodfellow, J. L. Finney & S. Neidle, *J. Biomol. Struct. Dyn.* **1**, 287–97 (1983).
38. V. Madison, D. S. Osguthorpe, P. Dauber & A. T. Hagler, *Biopolymers* **22**, 27–31 (1983).
39. P. L. Howell & J. M. Goodfellow, *J. Physique* **45**; *Suppl.* C7 211–18 (1984).
40. F. Vovelle, J. M. Goodfellow, H. F. Savage, P. Barnes & J. L. Finney, *Euro. J. Biophys.* **11**, 225–37 (1985).
41. J. L. Finney, J. M. Goodfellow, P. L. Howell & F. Vovelle, *J. Biomol. Struct. Dyn.*, **3**, 599–622 (1985).
42. W. C. Hamilton & J. A. Ibers, *Hydrogen Bonding in Solids*, Benjamin, New York, Amsterdam, 1968.
43. M. Falk and O. Knop, in *Water: A Comprehensive Treatise*, Vol. 2 (F. Franks, ed.), Plenum Press, New York, 1973, p. 55.
44. B. Pederson, *Acta Crystallogr.* B **30**, 289–91 (1974).
45. G. Ferraris & M. Franchini-Angela, *Acta Crystallogr.* B **28**, 3572–83 (1972).
46. G. Chiari & G. Ferraris, *Acta Crystallogr.* B **38**, 2331–41 (1982).
47. I. Olovsson and P. G. Jonsson, in *The Hydrogen Bond*, Vol. 2 (P. Schuster, G. Zundel & C. Sandorfy, eds), North-Holland, Amsterdam, 1976, p. 393.
48. E. Whalley, *J. Phys. Chem.* **87**, 4174–79 (1983).
49. A. Polian & M. Grimsditch, *Phys. Rev. Lett.* **52**, 1312–14 (1984).
50. K. S. Schweizer & F. H. Stillinger, *J. Chem. Phys.* **80**, 1230–40 (1984).
51. B. Kamb, in *Physics and Chemistry of Ice* (E. Whalley, S. J. Jones & L. W. Gold, eds.), Royal Society of Canada, Ottawa, 1973, p. 28.
52. G. P. Johari & E. Whalley, *J. Chem. Phys.* **70**, 2094–97 (1979).
53. W. F. Kuhs, J. L. Finney, C. Vettier & D. V. Bliss, *J. Chem. Phys.* **81**, 3612–23 (1984).
54. S. W. Peterson & H. A. Levy, *Acta Crystallogr.* **10**, 70–6 (1957).
55. W. F. Kuhs & M. S. Lehmann, *J. Chem. Phys.* **87**, 4312–13 (1983).
56. K. Shimaoka, *J. Phys. Soc. Japan* **15**, 106–19 (1960).
57. B. Kamb. *Acta Crystallogr.* **17**, 1437–49 (1964).
58. B. Kamb, W. C. Hamilton, S. J. LaPlaca & A. Prakash, *J. Chem. Phys.* **55**, 1934–45 (1971).
59. B. Kamb. & A. Prakash, *Acta Crystallogr*, B **24**, 1317–27 (1968).
60. S. J. LaPlaca, W. C. Hamilton, B. Kamb & A. Prakash, *J. Chem. Phys.* **58**, 567–80 (1973).
61. H. Engelhardt & B. Kamb. *J. Chem. Phys.* **75**, 5887–99 (1981).
62. B. Kamb, A. Prakash & C. Knobler, *Acta Crystallogr.* **22**, 706–15 (1967).
63. W. C. Hamilton, B. Kamb, S. J. LaPlaca & A. Prakash, in *Physics of Ice* (N. Riehl, B. Bullemer & H. Engelhardt, eds.), Plenum Press, New York, 1969, p. 44.
64. B. Kamb, *Science* **150**, 205–9 (1965).
65. J. D. Jorgensen, R. A. Beyerlein, N. Watenabe & T. G. Worlton, *J. Chem. Phys.* **81**, 3211–14 (1984).
66. B. Kamb & B. L. Davis, *Proc. Natl. Acad. Sci, USA* **52**, 1433–39 (1964).

67. J. D. Jorgensen and T. G. Worlton, *J. Chem. Phys.* **83**, 329–33 (1985).
68. G. H. F. Diercksen, *Theor. Chim. Acta* **21**, 335–67 (1971).
69. H. Popkie, H. Kistenmacher & E. Clementi, *J. Chem. Phys.* **59**, 1325–36 (1973).
70. J. A. Odutola & T. R. Dyke, *J. Chem. Phys.* **72**, 5062–70 (1980).
71. G. F. Diercksen, W. P. Kraemer & B. O. Roos, *Theor. Chim. Acta* **36**, 249–74 (1975).
72. H. Kistenmacher, G. C. Popkie & E. Clementi, *J. Chem. Phys.* **61**, 546–61 (1974).
73. O. Matsuoka, E. Clementi & M. Yoshimine, *J. Chem. Phys.* **64**, 1351–61 (1976).
74. J. O. Baum & J. L. Finney, *Mol. Phys.* **55**, 1097–108 (1985).
75. A. H. Narten, M. D. Danford & H. A. Levy, *Discuss. Faraday Soc.* **43**, 97–107 (1967).
76. J. C. Dore, *Water Science Reviews* **1**, 3–92 (1985).
77. W. E. Thiessen & A. H. Narten, *J. Chem. Phys.* **77**, 2625–62 (1982).
78. G. Palinkas, E. Kalman & P. Kovacs, *Mol. Phys.* **34**, 525–37 (1977).
79. P. Barnes, J. L. Finney, J. D. Nicholas & J. E. Quinn, *Nature* **282**, 459–64 (1979).
80. A. Beyer, A. Karpfen & P. Schuster, in *Topics in Chemistry 120: Hydrogen Bonds* (P. Schuster, ed.), Springer-Verlag, Berlin–Heidelberg–New York–Tokyo, 1984, p. 1.
81. A. C. Shepard, Y. Beers, G. P. Klein & L. S. Rothman, *J. Chem. Phys.* **59**, 2254–59 (1973).
82. E. Whalley, *J. Glaciol.* **21**, 13–31 (1978).
83. J. L. Finney & J. M. Goodfellow, in *Structure and Dynamics: Nucleic Acids and Proteins* (E. Clementi and R. H. Sarma, eds.), Adenine Press, New York, 1983, p. 81.
84. G. H. Stout & L. H. Jensen, *X-ray Structure Determination*, Macmillan, New York, 1968.
85. M. M. Woolfson, *An Introduction to X-ray Crystallography*, Cambridge University Press, Cambridge, 1970.
86. G. E. Bacon, *Neutron Diffraction*, Clarendon Press, Oxford, 1975.
87. T. L. Blundell & L. N. Johnson, *Protein Crystallography*, Academic Press, London, 1976.
88. K. D. Watenpaugh, T. N. Margulis, L. C. Sieker & L. H. Jensen, *J. Mol. Biol.* **122**, 175–90 (1978).
89. C. C. F. Blake, W. C. A. Pulford & P. J. Artymiuk, *J. Mol. Biol.* **167**, 693–723 (1983).
90. W. Baur, *Acta Crystallogr.* B **28**, 1456–65 (1972).
91. C. K. Nockolds, T. N. M. Waters, S. Ramaseshan, J. M. Waters & D. C. Hodgkin, *Proc. Indian Acad. Sci. (Chem. Sci.)* **93**, 195–234 (1984).
92. F. M. Moore, B. H. O'Connor, B. T. Willis & D. C. Hodgkin, *Proc. Indian Acad. Sci. (Chem. Sci.)* **93**, 235–60 (1984).
93. B. Klar, B. Hingerty & W. Saenger, *Acta Crystallogr.* B **36**, 1154–65 (1980).
94. K. Lindner & W. Saenger, *Carbohydrate Res.* **99**, 103–15 (1982).
95. C. Betzel, W. Saenger, B. E. Hingerty & G. M. Brown, *J. Am. Chem. Soc.* **106**, 7545–57 (1984).
96. H. F. J. Savage, *Ph.D Thesis, University of London*, 1983.

97. H. F. J. Savage, P. F. Lindley, J. L. Finney & J. Allibon, *Acta Crystallogr.*, submitted 1986.
98. H. F. J. Savage, *Biophys, J.*, part I, in press, 1986.
99. H. F. J. Savage, *Biophys, J.*, part II, in press, 1986.
100. H. S. Shieh, H. M. Berman, M. Dabrow & S. Neidle, *Nucleic Acids Res.* 8, 85–97 (1980).
101. S. Neidle, H. M. Berman & H. S. Shieh, *Nature* 288, 129–33 (1980).
102. I. L. Karle, and E. Duesler, *Proc. Natl. Acad. Sci. USA.* 74, 2602–6 (1977).
103. N. C. Seeman, R. O. Day & A. Rich, *Nature* 253, 324–6 (1975).
104. J. L. Finney, in *Water: A Comprehensive Treatise*, Vol. 6 (F. Franks, ed.), Plenum Press; New York, 1979, p. 47.
105. E. Baker & R. Hubbard, *Prog. Biophys. Mol. Biol.* 44, 97–179 (1984).
106. F. M. Moore, B. T. M. Willis & D. C. Hodgkin, *Nature* 214, 130–3 (1967).
107. P. G. Lenhert, *Proc. R. Soc. London A* 303, 45–84 (1968).
108. M. M. Teeter & A. A. Kossiakoff, in *Neutrons in Biology* (B. P. Schoenborn, ed.), Plenum Press, New York, 1984, p. 335.
109. S. A. Mason, G. A. Bentley & G. J. McIntyre, in *Neutrons in Biology* (B. P. Schoenborn, ed.), Plenum Press, New York, 1984, p. 323.
110. *Handbook of Chemistry and Physics*, *P.D–178*, Chemical Rubber Co, Cleveland, Ohio, 1972.
111. N. L. Allinger, *Adv. Phys. Org. Chem.* 13, 2–82 (1976).
112. B. Kamb, *J. Chem. Phys.* 43, 3917–24 (1965).
113. M. S. Lehmann, J. J. Verbist, W. C. Hamilton & T. F. Koetzle, *J. Chem. Soc. Perkin Trans. II*, 133–137 (1973).
114. G. E. Bacon, *Neutron Diffraction*, Clarendon Press, Oxford, 1975, p. 355.
115. W. F. Kuhs, *Acta Crystallogr. A* 39, 148–58 (1983).
116. W. F. Kuhs & M. S. Lehmann, in *Colston Symposium on Water and Aqueous Solutions, Bristol, 1985.*
117. C. N. R. Rao, in *Water: A Comprehensive Treatise*, Vol. 1. (F. Franks, ed.), Plenum Press, New York, 1972, p. 93.
118. G. A. Jeffery & R. K. McMullan, *Prog. Inorg. Chem.* 8, 43–108 (1967).
119. D. W. Davidson, in *Water: A Comprehensive Treatise*, Vol. 2. (F. Franks, ed.), Plenum Press, New York, 1973, p. 115.
120. R. D. Green, *Hydrogen Bonding by CH Groups*, Macmillan, London, 1974.
121. G. G. Dodson, in *Refinement of Protein Structures*, Daresbury Laboratory, SERC., UK, 1981.
122. K. Sakabe, N. Sakabe & K. Sasaki, in *Structural Studies on Molecules of Biological Interest* (G. Dodson, J. P. Glusker & D. Sayre, eds.), Clarendon Press, Oxford, 1981, p. 509.
123. W. R. Chang, D. Stuart, J. B. Dai, R. Todd, J. P. Zhang, D. L. Xie, B. Kuang & D. C. Liang. *13th IUCr. Congress Abstracts.* Hamburg, C16 (1984).
124. A. Wlodawer, H. F. J. Savage & G. G. Dodson, in preparation, 1986.
125. W. A. Hendrickson & M. M. Teeter, *Nature* 290, 107–13 (1981).
126. J. Walters & R. Huber, *J. Mol. Biol.* 167, 911–17 (1983).
127. A. Wlodawer, J. Walter, R. Huber & L. Sjölin, *J. Mol. Biol.* 180, 301–29 (1984).
128. I. Glover, I. Haneef, J. E. Pitts, S. P. Wood, D. Moss, I. Tickle & T. L. Blundell, *Biopolymers* 22, 293–304 (1983).

148   *Hugh Savage*

129. M. M. Teeter, *Proc. Natl. Acad. Sci. USA.* **81**, 6014–18 (1984).
130. J. Shpungin & A. A. Kossiakoff, in preparation (1985).
131. A. A. Kossiakoff, unpublished, 1985.
132. T. N. Bhat & D. M. Blow, *Acta Crystallogr.* A **38**, 21–9 (1982).
133. B. C. Wang, in *Methods in Enzymology* **115**, 90–112 (1986).
134. W. F. van Gunsteren, H. J. C. Berendsen, J. Hermans, W. G. J. Hol & J. P. M. Postma, *Proc. Natl. Acad. Sci.* **80**, 4315–19 (1983).
135. P. C. Moews & R. H. Kretsinger, *J. Mol. Biol.* **91**, 201–28 (1975).
136. M. S. Lehmann, S. A. Mason & G. J. McIntyre, *13th I.U.Cr. Congress Abstracts, Hamburg,* C53 (1984).
137. G. Kartha, J. Bello & D. Harker, *Nature* **213**, 862–5 (1967).
138. A. Wlodawer & L. Sjölin, *Biochemistry* **22**, 2720–8 (1983).
139. M. Fujinaga, L. T. J. Debaere, G. D. Brayer & M. N. G. James, *J. Mol. Biol.* **183**, 479–502 (1985).
140. G. G. Dodson & D. C. Hodgkin, personal communication.
141. F. Vovelle & M. Ptak, *Int. J. Pept. Protein Res.* **13**, 435–46 (1979).
142. M. D. Morse & S. A. Rice, *J. Chem. Phys.* **76**, 650–60 (1982).
143. P. Barnes, in *Progress in Liquid Physics* (C. A. Croxton, ed.), Wiley-Interscience, Chichester, 1978, p. 391.

The Nature of the Hydrated Proton. Part One: The Solid and Gaseous States

C. I. RATCLIFFE* AND D. E. IRISH†

* Division of Chemistry, National Research Council Canada, Ottawa, ON, Canada K1A 0R6
† Department of Chemistry, University of Waterloo, Waterloo, ON, Canada N2L 3G1

'Ions are jolly little beggars', Rutherford used to say.
'You can almost see them' Eve [1].

These words, attributed to Rutherford, provide a focus for this article. To what extent can one 'see' that most important of all the cations – the hydrated proton? What do different eyes – different experimental techniques – see and how do the views from these differing points of observation differ or complement one another? Our search will encompass the three states of matter – solid, gas and liquid solutions. And our 'eyes' will include X-ray and neutron diffraction, nuclear magnetic resonance, pulsed-electron-beam high-pressure mass spectrometery, infrared and Raman spectroscopy and relaxation techniques.

In addition we must consider the 'mind's eye': quantum simulations and theoretical inferences. But first we will place our search in historical perspective.

1. Historical Perspective [2–5]

The recognition that the class of compounds generally recognized as acids had, in common, the element hydrogen can be attributed to Davy and Liebig. In 1838 Liebig defined acids as 'compounds containing hydrogen, in which hydrogen can be replaced by a metal' [2]. This concept unified a body of empirical knowledge that had been accumulating: the sour taste; the solvent power; the ability to turn blue vegetable dyes red; the ability to combine with bases (or alkalis) to form salts and water; the existence of acidic compounds which did not contain oxygen, e.g. HF, HCl, HBr etc. (a discovery by Davy (*circa* 1810) that laid to rest the oxygen theory of acids proposed by Lavoisier).

The recognition that the acid property was to be associated with those labile hydrogen atoms which could produce hydrogen ions in aqueous solutions came from the electrolytic dissociation theory of Arrhenius and Ostwald (1880–90).

'Electrolytes are salts in aqueous solution. The term "salt" is here used to include both acids and bases, in that acids are salts of hydrogen, and bases salts of hydroxyl' Wm Ostwald (1894) [2].

The strength of an acid could reasonably be expressed by its dissociation constant, K_a. The neutralization reaction was then expressed as

$$H^+ + OH^- \rightarrow H_2O$$

concurrently providing the 'spectator' ions of a salt which could be recovered by evaporation of the solvent. Such a formulation relegated the role of the solvent to that of an inert medium in which the physical chemistry takes place. It also failed to adequately account for acid-base reactions in non-aqueous solvents. But even as early as 1886 H. E. Armstrong noted that binary hydrogen compounds in the liquid state were not electrolytes *per se*. Water was essential. A. Werner (1913) extended this theme with introduction of the terms anhydro- and aquo-acids: the former are compounds which combine with the hydroxyl ions of water to thus establish their characteristic hydrogen ion concentration; the latter add to water with subsequent dissociation to yield hydrogen ions [2].

The identification and characterization of the sub-atomic particle, named the proton by Rutherford, [6] resulted from the α-ray scattering experiments of Rutherford and later by Blackett and the study of positive rays by Thompson, Goldstein, Wien, Dempster, Aston and others [7]. This particle has a rest mass of $1.6726485\,(86) \times 10^{-27}$ kg, a positive charge of $1.6021892\,(46) \times 10^{-19}$ C and a diameter of 1.2×10^{-3} pm – some 3×10^4 smaller than the covalent radius of the hydrogen atom (37 pm) [8, 9]. It is reasonable to assume that this particle of such small size and high charge density will be intimately associated with the electron density of neighbouring atoms in condensed phases and that the free proton is not a plentiful species.

The definition of acids and bases proposed by Brönsted [10] and Lowry [11] in 1923 remains as one of the most useful: An acid is a species having a tendency to lose a proton, and a base is a species having a tendency to add a proton. The scheme $A = B + H^+$ introduces the concept of conjugate (or corresponding) acid–base pairs, which differ from each other by one proton. A and B may themselves be charged. This definition is independent of the solvent, in that H^+ is a proton. However the application of the definition to typical acid–base reactions does introduce the solvent as an active participant, B_2. Thus

$$A_1 + B_2 = A_2 + B_1$$

involves two conjugate acid–base pairs A_1–B_1 and A_2–B_2. When B_2 is H_2O, A_2 is the H_3O^+ ion; when B_2 is NH_3, A_2 is NH_4^+; and so on. This scheme

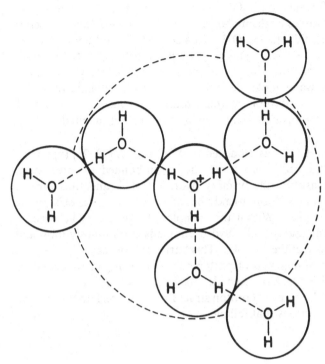

Figure 1. Hydration model of the proton in aqueous solution [23].

has the virtue that a ranking of conjugate acid–base strengths follows logically [12]. It does however, beg the question of the degree of hydration of the proton.

The broadened concept of acid as an electron-pair acceptor by Lewis [4, 13] is advantageous in some areas of chemistry but is not directly related to our more restricted theme, except that the proton is indeed an electron-pair acceptor. Bjerrum has proposed that Lewis 'acids' might better be called antibases, thus reserving the name acid for the protonic acids [14].

It was a natural consequence of the Brönsted theory that the hydrogen ion, in aqueous solution, should be written as the oxonium ion, H_3O^+. Volmer [15] first suggested that H_3O^+ existed in a crystal lattice; he noted the similarity in the X-ray powder patterns of NH_4ClO_4 and $HClO_4 . H_2O$ and surmised that the latter was completely ionized as $H_3O^+ . ClO_4^-$. This was confirmed by two independent proton magnetic resonance studies [16, 17]. The solid acid hydrates $HA . H_2O$ $(A = NO_3^-,\ ClO_4^-,\ HSO_4^-)$ were found to be ionic, $H_3O^+A^-$, and H_3O^+ was found to have a flattened, pyramidal shape. Prior to this, Goldschmidt and Udby [18] had attributed the catalytic action in ester formation in aqueous solutions to the complex $(H_2O, H)^+$. Hantzsch [19] reached similar conclusions and introduced the term hydroxonium ion or the shorter term hydronium ion for H_3O^+.

Support for the formulation H_3O^+ was provided by Fajans and Joos [20], who correlated the values of the molar refractions of the isoelectronic series O^{2-} (6.9), OH^- (5.1), OH_2 (3.75), OH_3^+ (3.04 cm^3 mol^{-1}) and noted that H^+ would have a negative value. Fajans [21] had previously estimated from an electrostatic picture that 837 kJ mol^{-1} of energy would be released on union of a proton with a water molecule; this gives a crude estimate of 10^{-150} mol dm^{-3} for the concentration of free protons in aqueous solutions of acids at ambient temperatures [3], again pointing to a predominant hydrated species.

The hydration of H_3O^+ was modelled by Wicke *et al.* [22, 23] (figure 1). The H_3O^+ ion in the centre is strongly hydrogen-bonded to three water molecules which constitute the primary hydration shell and thus form a pyramidal $H_9O_4^+$ complex. More weakly bound water molecules exist in the secondary hydration shell. Within this complex the proton is believed to oscillate continuously and quickly; hydrogen bonds continuously form and break at the periphery of the complex. This 'structural diffusion' was stated to be the rate-determining step of protonic motion in liquid water. Other hydrates, $H_7O_3^+$ and $H_5O_2^+$, have been identified in crystals. It is this 'variably hydrated' aspect of the hydrated proton structure that we seek to understand more clearly in the sections that follow.

2. Nomenclature

The opinion of the IUPAC Commission [24] is:

'3.14   Names for polyatomic cations derived by addition of more protons than required to give a neutral unit to monoatomic anions, are formed by adding the ending – onium to the root of the name of the anion element.'

Example: oxonium for H_3O^+

'The ion H_3O^+, which is in fact the monohydrated proton, is to be known as the oxonium ion when it is believed to have this constitution, as for example in $H_3O^+ClO_4^-$, oxonium perchlorate. If the hydration is of no particular importance to the matter under consideration, the simpler term hydrogen ion may be used. The latter may also be used for the indefinitely solvated proton in non-aqueous solvents; but definite ions such as $CH_3OH_2^+$ and $(CH_3)_2OH^+$ should be named as derivatives of the oxonium ion, i.e. as methyl and dimethyl oxonium ions respectively.'

'The committees concur in oxonium for the ion H_3O^+, but see little reason for encouraging retention of the term hydronium ion because hydrogen ion adequately designates an intermediate degree of hydration' [25].

Two other terms are convenient. An *ionogen* is a substance with a molecular crystal lattice which can produce an electrolyte by reaction when dissolved in appropriate solvents. Thus the substances HCl, HNO_3 and $HClO_4$ are ionogens. An *ionophore* is a substance in which ions and only ions are present in the crystal lattice. Thus NaCl exists as Na^+Cl^- and $HClO_4 . H_2O$ exists as $H_3O^+ . ClO_4^-$; these are ionophores. These terms (and others) are discussed by Fuoss [26]. If in a particular solvent at a specified temperature the equilibrium state of an acid favours a high population of ionogen the acid is said to be *weak*; if the population of ionogen is low or zero the acid is *strong* (e.g. 1 M $HClO_4$ and 1 M HCl in water at 25 °C are strong ionogens).

3. The Solid State

It is in the solid state where the hydrated proton has best been characterized since (by the very nature of this state) the atoms are located at specific lattice sites; even so in some cases static or dynamic disorder can conceal information. Studies of the solid state provide answers to two major questions: (1) Where is the excess acid proton? (2) How does it interact with its environment and in particular what effects does it have on adjacent water molecules? Such information for solids may thus prove useful in gaining more insight into the nature of the hydrated proton in solution.

Several reviews of various aspects of the hydrated proton in solids have appeared in the past [27–37] and here the reader's attention is drawn particularly to the excellent reviews of structural studies by Lundgren and Olovsson [29, 33] and Taesler [31] and of vibrational spectroscopic studies by Williams [34]. It would be pointless to repeat the in-depth discussions of these reviews and in the following we emphasize the major points, give some specific examples and extend the review to cover all of the more recent work.

A collection of references to studies pertaining to all the known stoichiometric solids (and a few non-stoichiometric types) containing the hydrated proton is given in the appendix. Inevitably there will be some omissions (apologies offered); papers have been omitted which refer only to preparations of materials which have later been extensively characterized.

The majority of known oxonium-containing compounds are inorganic acids or acid salts, though studies dating from 1964 have revealed that a number of strong organic acid hydrates (frequently involving an aromatic ring and a sulphonic acid substituent) also contain the oxonium species. One distinct but important category, which has usually been overlooked in previous reviews, is oxonium-containing minerals [84, 243–266]. From a historical perspective the presence of H_3O^+ had long been suspected in many minerals, a view based mainly on elemental compositional analysis and IR spectroscopy. Evidence has since confirmed the presence of oxonium ions in a number of these materials [84, 259–261]. Frequently the oxonium ion occurs as a substitute for K^+ ions and the degree of substitution may vary from 0 to 100 per cent.

The important role played by the determination of acid/H_2O composition *versus* freezing point curves in discovering numerous hydrates of the strong inorganic acids [305–312], whose true ionic nature was determined only later, is noteworthy. Even in recent years such studies have preceded the preparation and characterization of new ionized acid hydrates [38, 120, 127, 189, 267].

3.1. *Structural Studies*

X-ray and neutron diffraction structural studies of solids provide extensive information on the hydrated proton. They also show that by no means do all acid hydrates form ionic solids, e.g. $(COOH)_2 . 2H_2O$ [313], $(H_3PO_4)_2 . H_2O$ [314] and $NaHSO_4 . H_2O$ [315]. Another example is $BF_3 . 2H_2O$ which is partially ionized to H_3O^+ and BF_3OH^- in its liquid state [316] but is definitely not ionized in the solid [317]. Nor is it possible to correlate the solid form (ionophore or ionogen) of weak or medium strength acids with their dissociation constants. For instance $CF_3COOH . H_2O$ is a true molecular acid hydrate (ionogen) whereas $CF_3COOH . 4H_2O$ is an ionic salt (ionophore) [267]. Also, as pointed out by Taesler [31], $H_2SO_4 . 2H_2O$ forms $(H_3O^+)_2SO_4^{2-}$ rather than $(H_5O_2^+) (HSO_4^-)$ [117], even though HSO_4^- has a higher pK_a value than oxalic acid which is not ionized in its dihydrate solid [313].

One must bear in mind the quality of the X-ray or neutron diffraction structural determination. For example the results of the first X-ray studies on the monohydrates of nitric [104] and sulphuric [112] acids were interpreted in terms of associated acid species. Later more accurate determinations by Taesler [105, 113] showed them to be ionic oxonium salts. It is possible that other oxonium solids have not been recognized as such; an early study of an organic acid hydrate, i.e. dilituric acid [296], shows O–O distances which may indicate that the material is ionized. Another complication occurs in structures which show disorder of the proton and/or oxygen positions; there are numerous cases (for example [63, 132, 138, 140, 146, 199, 234, 235, 273, 279, 304]). In $Y(C_2O_4)_2(H_5O_2) . H_2O$ there is a clearly dynamic disorder, since the disorder increases with temperature [138]. There are also cases where the determination of the structure is made ambiguous on account of possible disorder [304]. In one instance the disorder of an O atom indicates an associated acid in one position or the dissociated acid with $H_5O_2^+$ in the other position [132].

It is clear from all the reliable structures that the excess acid proton is bound to a water molecule either in isolation or in association with one or more other water molecules via hydrogen bonds. There are different ways of classifying these higher hydrates. Lundgren and Olovsson [29, 33] and Taesler [31] all conclude that the basic units are H_3O^+ and $H_5O_2^+$ and that all higher hydrate formulations such as $H_7O_3^+$ and $H_9O_4^+$ are artificial but convenient labels for what are really $H_3O^+ . nH_2O$ or $H_5O_2^+ . nH_2O$. Even $H_5O_2^+$ is in some cases better described as $H_3O^+ . H_2O$. In fact the boundary between $H_5O_2^+$ and $H_3O^+ . H_2O$ is not well defined. We suggest, however, that $H_5O_2^+$ be reserved for species in which the following two criteria are met:

(1) The O–H distance be greater than or equal to 110 pm, and

(2) The O...O distance be less than or equal to 248 pm.

These two values correspond to a point on the well known O...O *versus* O–H plot [318, 319 and Savage, this volume, p. 67]. These limits are of course still somewhat artificial but they do give recognition to the uniqueness of those cases which possess such a short and strong hydrogen bond, and which might be expected to have properties different from H_3O^+ attached by a weaker hydrogen bond to H_2O. If the structural information is detailed enough to answer the question 'where is the excess proton?', it is then always possible to describe the hydrated proton in terms of $H_3O^+ . nH_2O$ or $H_5O_2^+ . nH_2O$. Neutron diffraction studies are the most useful in this respect since they often allow precise and unambiguous determination of the H or D atom positions.

X-ray studies form the large majority of the structural determinations. While the positions of the H atoms may not be determined with any great accuracy, if at all, it is sometimes possible to unambiguously locate three H atoms around one O atom. When H atoms are not determined, however, interatomic distances and orientations become the key to the form of the hydrated proton, and it is here where it is often more convenient to use such labels as $H_7O_3^+$ etc. for higher hydrated types. Taesler [31] states that one criterion used to distinguish such units within even larger aggregates of H_2O molecules is that they be held together by hydrogen bonds which are more than about 20 pm shorter than hydrogen bonds to the other water molecules. Two or three roughly equivalent short hydrogen bonds to H_2O molecules around one oxygen would be taken as indicative of $H_7O_3^+$ or $H_9O_4^+$ respectively. Such species labels, as given by the respective authors, are given in the appendix. It should be noted that the geometry of the higher hydrated forms shows a great deal of variation from one structure to another.

In cases with only one or two water molecules per proton the H_3O^+ or $H_5O_2^+$ groups are always isolated. Higher hydrates, however, will form either isolated groupings or interlink to form chain (for example [50–52, 111, 122, 128, 199]), layer (for example [98, 129, 151, 165, 166, 198, 295]), or three-dimensional (for example [55, 99, 102]) structures. Some rare cases with quite symmetrical structures have been labelled as $H_{14}O_6^{2+}$ [209] and $H_{13}O_6^+$ [303]. Isolated aggregates occur in general when the anions are large (for example, most of the substituted aromatic organo-sulphonic acid higher hydrates [272, 275, 284, 290, 292]) and in one instance where there is also a large cation present [303]. Smaller anions generally cause extended hydrogen-bonded structures to form and many of the water molecules will also be

Table 1. *Average hydrogen-bond lengths (pm) in hydrated proton species in solids*

(1) Isolated H_3O^+	
Ordered: 260.9 (8.6)*	Disordered: 287.3 (5.2)

(2) Species containing H_3O^+ connected to other H_2O

$H_3O^+ \cdots (H_2O)_p$: 255.7 (9.2)	$H_3O^+ \cdots Ox$: 268.7 (13.2)
$(H_2O)_p \cdots (H_2O)_s$: 276.1 (11.1)	$(H_2O)_p \cdots Ox$: 283.5 (8.7)
$H_3O^+ \cdots Cl^-$: 293.5 (1.6)	$H_3O^+ \cdots F^-$: 250.3 (7.5)
$H_3O^+ \cdots Br^-$: 308.5 (4.9)	$H_3O^+ \cdots (XF_n^{m-})$: 262.4 (9.1)

(3) $H_5O_2^+$ containing species

$O-H' \cdots O$: 242.4 (2.6) (cases $\leqslant$ 2.48)	245.6 (6.9) (all claimed cases)
$H_5O_2^+ \cdots (H_2O)_p$: 270.2 (5.6)	$H_5O_2^+ \cdots Ox$: 270.2 (6.0)
$H_5O_2^+ \cdots Cl^-$: 304.3 (5.6)	$H_5O_2^+ \cdots Br^-$: 323.3 (2.9)
$H_5O_2^+ \cdots (XCl_n^{m-})$: 315.4 (0.3)	$H_5O_2^+ \cdots BF_4^-$: 275.3 (10.6)
	$H_5O_2^+ \cdots CN^-$: 273.0 (2.3)

* Values in parentheses are standard deviations.
Subscripts $_p$ and $_s$ on (H_2O) signify primary and secondary water respectively.
Ox signifies an oxyanion.
XF_n^{m-} signifies a polyfluoro anion.
XCl_n^{m-} signifies a polychloro anion.

involved in the coordination sphere of the anion (for example [50–52, 55, 111]). There are exceptions to these generalizations and the anions obviously have considerable influence on the hydrogen-bonded network of the hydrated proton. One might emphasize that there are only a few cases in which H_2O molecules co-exist entirely separately from H_3O^+ or $H_5O_2^+$ within the same structure, (for example [136, 138, 141–143, 145, 195, 259]). All of these cases contain a metal atom (Er, Y, Cr, In, V, Mo, Fe) to which the separate H_2O molecules are strongly coordinated via the O atom, so that the H_2O molecules would be unable to accept hydrogen bonds from the hydrated proton species.

Another more unusual type of hydrated proton occurs in the clathrate hydrate species $HXF_6 . HF . 4H_2O$ (X = P, As, Sb) [234, 235]. Here the XF_6^- ions are enclathrated in cages built up from O and F atoms (at the cage vertices) held together by hydrogen bonds. The H atoms have not been located but are presumed to be disordered (as are the O and F atoms). Mootz [120] has recently determined structures of compounds of composition $HPF_6 . 7 . 7H_2O$, $HBF_4 . 5 . 5H_2O$ and $HClO_4 . 5 . H_2O$ which are related to structure-I-type gas hydrate clathrates, and $HXF_6 . 6H_2O$ (X = As or Sb) related to structure-IV hydrate cathrates [320].

Table 1 gives averaged values of hydrogen-bond lengths taken from all known structures, of reasonable accuracy, for H_3O^+ and $H_5O_2^+$ in situations either isolated from, or attached to other water molecules. A few important

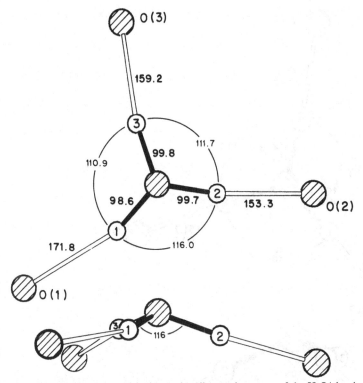

Figure 2. The geometry and hydrogen-bonding environment of the H_3O^+ ion in $H_3O^+CF_3SO_3^-$ from neutron diffraction results, replotted from Lundgren *et al.* [126]: Views parallel and perpendicular to the plane of the three H atoms.

generalizations can be made based on these averages which go some way to answering the question 'how does the hydrated proton affect its environment?': (1) Both H_3O^+ and $H_5O_2^+$ prefer to hydrogen bond via H donation to the anions or to other waters; (2) hydrogen bonds to neighbouring (primary) water molecules are usually the shortest; (3) hydrogen bonds from primary waters to secondary waters, i.e. those not attached to H_3O^+ or $H_5O_2^+$, are still shorter than the H_2O–H_2O hydrogen bonds in ice (276 pm) [321]; (4) hydrogen bonds from anions directly to H_3O^+ or $H_5O_2^+$ are generally shorter than from anions to primary water molecules. Clearly the positive charge enhances hydrogen bonding. Also, judging by the distances involved, there appears to be a greater affinity for hydrogen-bond formation with water O atoms than anionic O atoms. We will now consider the basic units H_3O^+ and $H_5O_2^+$ in more detail.

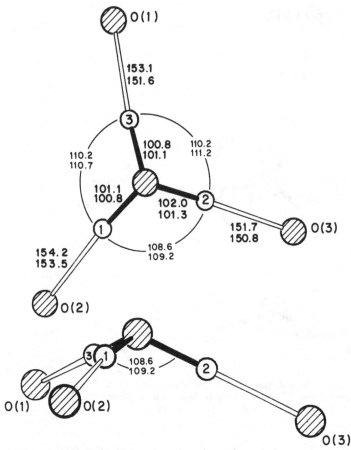

Figure 3. The geometry and hydrogen-bonding environment of the H_3O^+ (D_3O^+) ion in H_3O^+para-$CH_3(C_6H_4)SO_3^-$ from neutron diffraction results, replotted from Lundgren and Williams [279] and Finholt and Williams [280]: views parallel and perpendicular to the plane of the three H(D) atoms. Parameters from the D_3O^+ determination [280] are given above the values for the H_3O^+ determination [279]. Note the much less distorted hydrogen-bonding arrangement compared with $H_3O^+CF_3SO_3^-$ in figure 2.

3.2. The Structure of H_3O^+

In the great majority of substances containing isolated H_3O^+ the three hydrogens each form a hydrogen bond to a neighbouring anion, so that the O atoms and the three bonds form a well-defined pyramidal equilibrium configuration. In several cases there is even three-fold symmetry [42, 52, 58, 59]. There are, however, a few cases where this does not occur and the environment suggests that the H_3O^+ is orientationally disordered. Curiously, the first characterized oxonium salt, i.e. the room temperature phase of

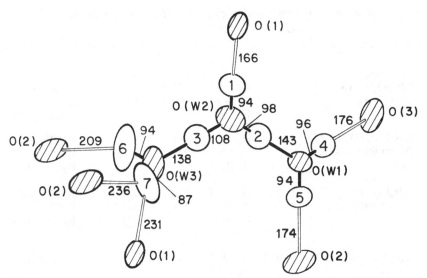

Figure 4. The geometry and hydrogen-bonding environment of $H_7O_3^+$ in 2,5-$Cl_2(C_6H_3)$ $SO_3H . 3H_2O$ from neutron diffraction results replotted from Rozière and Williams [273]: The isolated $H_7O_3^+$ unit can clearly be described as $H_3O^+ . 2H_2O$.

$H_3O^+ClO_4^-$, is one of these atypical cases. Neutron diffraction studies have confirmed the disorder [63, 64]. Recent NMR studies, to be discussed later, also indicate orientational disorder in the oxonium-substituted alunite-jarosite minerals [249], and the same situation probably holds in the structurally similar oxonium gallium sulphate [146]. These materials have significantly greater O···O distances from the oxonium to the oxyanions (values range from 282 to 299 pm) [60, 61, 146, 322, 323], indicating weaker hydrogen-bonding interactions. Note in contrast that the low-temperature phase of $H_3O^+ClO_4^-$ [62] has a more typical H_3O^+ ion with three O–H···O hydrogen bonds (average length 266 pm).

Surprisingly, very few oxonium-containing materials have been studied with high precision using neutron diffraction; $CH_3C_6H_4SO_3^-H_3O^+$ [279, 280] and $CF_3SO_3^-H_3O^+$ [126] are the only examples with isolated H_3O^+ ions (See figures 2 and 3). The H atom positions determined in the neutron studies of the $H_7O_3^+$ salts of $Cl_2C_6H_3SO_3^-$ [273] (see figure 4) and $Br_2C_6H_3SO_3^-$ [276] show that these really contain $H_3O^+ . 2H_2O$; however, their structures are complicated by the possibility of disorder which affects the H_3O^+ geometry. Parameters from selected neutron diffraction determinations are given in table 2.

In recent years structural studies have appeared of the rather more unusual type of H_3O^+ found in crown ether complexes. These salts generally consist of a monohydrated acid with the H_3O^+ complexed by the crown ether. In the two solids studied by X-ray diffraction the H_3O^+ is complexed by 18-crown-6

Table 2. *Bond distances (pm) and angles (°) in H_3O^+ ions from selected neutron diffraction studies*

		O–H	O–H$\cdots$O	∡ O–H$\cdots$O	∡ H–O–H
Isolated species					
$CF_3SO_3H.H_2O$	H_1	98.6	267.3	162.0	H_1OH_2 116.0
21 °C [126]	H_2	99.7	252.2	170.7	H_2OH_3 111.7
	H_3	99.8	257.9	169.2	H_3OH_1 110.9
$CH_3C_6H_4SO_3H.H_2O$	H_1	101.1	254.5	172.3	H_1OH_2 109.2
Room temperature	H_2	101.4	252.2	178.0	H_2OH_3 111.2
[280]	H_3	101.3	252.9	174.3	H_3OH_1 110.7
$CH_3C_6H_4SO_3D.D_2O$	H_1	101.1	254.9	173.1	H_1OH_2 108.6
Room temperature	H_2	102.0	253.7	177.4	H_2OH_3 110.2
[280]	H_3	100.8	253.7	174.8	H_3OH_1 110.2
Attached to H_2O					
$C_6H_3COOH(OH)SO_3H.3H_2O$	H_1	101.8	—	—	112.1
[31, 290]	H_2	109.5	—	—	114.9
	H_3	99.0	—	—	113.8
$C_9H_4NO_4H.2H_2O$	H_1	100.8	256.3	160.9	H_1OH_2 112.2
[301]	H_2	99.7	264.3	177.2	H_2OH_3 111.8
originally formulated	H_3	108.7	247.7	179.2	H_3OH_1 113.1
as $H_5O_2^+$					

derivatives [238, 239]. The H_3O^+ ion was found to be hydrogen-bonded to three of the six O atoms which are part of the cyclic crown.

3.3. The Structure of $H_5O_2^+$

If the two criteria proposed earlier to define this species are used then we can only be certain of about ten examples, relying mainly on neutron diffraction studies. However, if those X-ray structures which fulfil the O$\cdots$O bond-length requirement, but whose O–H distances are unknown, are considered also, the number increases substantially. Only a very few of the previously claimed $H_5O_2^+$ species must then be excluded [122, 198, 199, 206, 233, 293, 301]. A very few cases are ambiguous on account of insufficient details or disorder.

Table 3 gives parameters from neutron diffraction studies of $H_5O_2^+$. The most characteristic feature of this ion is its short central O–H'$\cdots$O hydrogen bond (tables 1, 3) which is either linear or close to linear (within 9°) in which the proton may be centred, disordered among two positions on either side of the centre (i.e. as a requirement of the crystal symmetry), or simply off-centre. The criterion requires that the O–H' bond be greater than or equal to 110 pm, which is considerably longer than the peripheral O–H bonds which average 98.2 pm. There is usually a retention of the flattened pyramidal geometry found in H_3O^+, around both of the oxygen atoms, and hydrogen bonds with external atoms are only donated, not accepted. In fact an accepted hydrogen bond from an external atom is sometimes used to argue against a formulation in terms of $H_5O_2^+$ [98]. The structure of $H_5O_2^+$ seems to be very dependent on the environment, which also has a strong effect on the position of the central H'.

It is clear that this ion has a good deal of conformational flexibility with respect to the relative angular positions around the O–H–O axis of the two end H_2O units. (This is also predicted by *ab initio* calculations [324] discussed in Part Two (to be published in *Water Science Reviews*, Vol. 3.) Williams categorized three types [28];

    (a) *cis* (boat), linear O–H'$\cdots$O, point symmetry mm (C_{2v});

    (b) *trans* (chair), linear O–H'$\cdots$O, point symmetry 2/m (C_{2h});

    (c) *gauche*, non-linear O–H'$\cdots$O, point symmetry 1 (C_1).

Obviously with this degree of variation the internal steric barriers to rotation of the H_2O end units must be relatively small compared to the external hydrogen-bond energies, which thus largely determine the conformation.

Another interesting point to consider is that in almost all cases where thermal ellipsoids have been determined for the central proton from neutron diffraction [53, 138, 145, 171, 284, 288, 295], the ellipsoid is elongated more or less in the direction of the bond. In centro-symmetric cases [138, 145, 171, 288] this has been taken to mean that there is either static or dynamic disorder of the proton between two sites. In only one centro-symmetric case [145] has a virtually isotropic thermal ellipsoid been found for the central proton (see figure 5). Thus this is possibly the only known centred O$\cdots$H'$\cdots$O bond in $H_5O_2^+$. (Some purists who do not like the criteria applied to define $H_5O_2^+$ might even argue that this is the only true example of $H_5O_2^+$.)

3.4. *Vibrational Spectroscopy*

As mentioned earlier, there is already in existence an excellent review by Williams on the vibrational spectroscopy of the hydrated proton [34]. In the intervening years there have been numerous papers which have reported vibrational frequencies for hydrated proton species, but the overall view has changed little.

Vibrational spectroscopy can provide indirect evidence of the hydrated proton in solids by showing the presence of bands due to the anion rather than those of the ionogen [66]. Indeed, in some cases this may be the only concrete evidence yet available to show that the material is an oxonium- or hydrogen ion-containing species, (e.g. $(H^+)_m(H_2O)_x \cdot MX_n{}^{m-}$ types where M is a metal, X = F, Cl, Br, n = 5, 6 and m = 1, 2 [172, 182–186, 189, 190, 193, 207, 211]).

3.4.1. *Spectra of H_3O^+*. For a C_{3v} H_3O^+ ion the spectrum should, in principle,

Table 3. Bond distances (pm) and angles (deg) in $H_5O_2^+$ ions from selected neutron diffraction studies

Compound	O_1-H'-O_2	O_1-H-H'	O_2-H'	$\not\prec O_1$-H'-O_2	$\not\prec$ H-O_1-H‡	$\not\prec$ H-O_2-H‡
$C_6H_2(NO_2)_3SO_3H \cdot 4H_2O$ [295]	243.6	112.8	131.0	175.0	115.4 110.4 111.9	112.6 107.2 108.1
$C_6H_4(COOH)SO_3H \cdot 3H_2O$ [284]	241.4	120.1	121.9	172.5	115.7 107.6 113.1	116.1 109.0 115.3
$C_6H_3(COOD)(OD)SO_3D \cdot 2D_2O$ [288]	(a) 243.5	121.7	†	180.0	114.8 112.0 114.5	— — —
	(b) 244.2	122.1	†	180.0	111.3 105.7 112.4	— — —
$Co(en)_2Cl_3 \cdot HCl \cdot 2H_2O$ [171]	243.1	121.6	†	180.0	114.5 109.1 110.5	— — —
$V(H_2O)_6(CF_3SO_3)_4H \cdot 2H_2O$ [145]	243.0	121.5	†	180.0	108.4 106.8 113.7	— — —

Compound	O_1–H	∡ O_1HO_x	$O_1 \cdots O_x$	O_2–H	∡ O_2HO_x	$O_2 \cdots O_x$
$C_6H_2(NO_2)_3SO_3H \cdot 4H_2O$ [295]	99.8	176.3	262.1	97.6	166.3	273.4
	99.6	177.2	260.4	97.7	168.7	268.5
$C_6H_4(COOH)SO_3H \cdot 3H_2O$ [284]	98.0	170.6	268.5	98.4	173.7	272.1 $\Big\}$ Bifurcated
	99.4	176.8	267.1	95.6	—	272.4
						280.2
$C_6H_3(COOD)(OD)SO_3D \cdot 2D_2O$ [288]	(a) 97.9	168.1	272.4	—	—	—
	97.9	165.3	265.8	—	—	—
	(b) 97.7	179.1	260.3	—	—	—
	98.0	168.1	275.5	—	—	—
$Co(en)_2Cl_3 \cdot HCl \cdot 2H_2O$ [171]	99.5	174.6	299.9 (O $\cdots$ Cl)	—	—	—
	98.8	177.9	303.4 (O $\cdots$ Cl)	—	—	—
$V(H_2O)_6(CF_3SO_3)_4H \cdot 2H_2O$ [145]	97.5	165.5	270.2	—	—	—
	97.1	170.9	278.0	—	—	—

† In centro-symmetric cases H' has been placed at the O–H'–O bond mid-point.

‡ 'H₂O' bond angle is underlined, H' is the central H atom, O_x is an external O atom (of the anion).

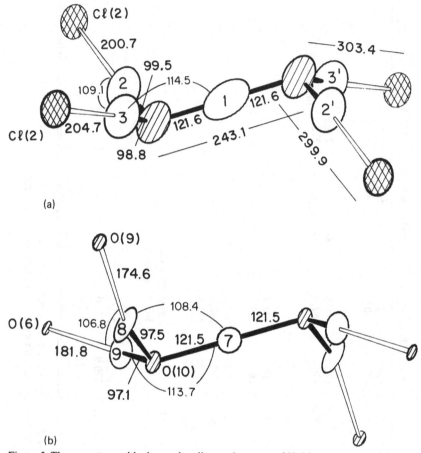

(a)

(b)

Figure 5. The geometry and hydrogen-bonding environment of $H_5O_2^+$ in *trans*-[Co(en)$_2$Cl$_2$]Cl . H$_5$O$_2$Cl [171] (a) and [V(H$_2$O)$_6$][H$_5$O$_2$][CF$_3$SO$_3$]$_4$ [145] (b) from neutron diffraction results. Compare in particular the elongated (a) and almost spherical (b) thermal ellipsoids of the central H′ in these two salts.

be quite simple and similar to that of NH_3 [65, 325]. The four normal modes are given in table 4. In the gas phase these modes are doubled because of inversion (discussed later) but this is not likely to be encountered in the solid since the hydrogen bonding and other steric forces should cause a dramatic increase in the inversion barrier for practically all cases. In solids, however, there are other factors which complicate the spectra:

The ion is subject both to the crystal site symmetry, which may lift the degeneracy of the C_{3V} (E) modes (see table 4), and to the factor group symmetry, which may increase the number of modes according to the number of ions in the primitive cell and the space group of the crystal. The bands of H_3O^+ may be overlapped by those of the anion; obviously this is not a

Table 4. *Normal mode descriptions and symmetries for* H_3O^+ *under the different possible molecular point groups*

Mode description		C_{3v}	C_s	C_1
Symmetric stretch	ν_1	A_1	A'	A
Symmetric bend	ν_2	A_1	A'	A
Anti-symmetric stretch	ν_3	E	$A'+A''$	2A
Anti-symmetric stretch	ν_4	E	$A'+A''$	2A

All modes are IR and Raman active.
Normal modes for C_{3v} are illustrated in Nakamoto [325] and for C_s in Basile *et al.* [281].

problem for monatomic anions, or polyatomic anions whose modes are all below approximately 900 cm⁻¹. The region of the O–H stretching modes may be complicated by overtones and combination bands, with the added possibility of Fermi resonance between modes of like symmetry.

If the spectral bands were sharp and well resolved most of these problems could be overcome, but the task is made most difficult by the largest effect involved, i.e. hydrogen bonding. The effects of hydrogen bonding on spectra in many types of compounds are well known [326]; it causes a broadening of bands (which is particularly dramatic in the case of the O–H stretching modes), uneven intensity increases, and shifts in frequency. There is a well known correlation for O–H $\cdots$ O hydrogen bonds between decreasing ν(O–H) frequencies and decreasing hydrogen-bond length [327, 328] (i.e. equivalent to increasing hydrogen-bond strength). It has been suggested that the bending mode frequencies increase with increasing hydrogen-bond strength.

Thus the wide range of different O–H $\cdots$ X bonds combined with differing symmetries in H_3O^+ solids not only makes the assignments of the fundamentals difficult, but also results in there being no specific 'characteristic' frequencies which can be used to identify the ion in any solid; instead broad ranges of frequencies exist. Each case must be considered separately. Table 5 gives the ranges of quoted peak values for the modes from all available studies. In compiling this table the difficulty experienced by authors in making assignments became very apparent, since the values quoted in different studies of the same material often differ by about 150 cm⁻¹, and in one case for the ν(O–H) assignment the difference is about 720 cm⁻¹. Considering their halfwidths and shapes (see also table 5) it would be difficult in some cases to justify quoting specific ν(O–H) band frequencies.

ν_1 and ν_3 have not been distinguished in table 5 since there is some controversy over their order. In NH_3 the order is $\nu_3 > \nu_1$ [325] and a number of authors have made a similar assignment for H_3O^+ by analogy. Intensity arguments lead to assignments in some cases of $\nu_3 > \nu_1$ [187] and in other cases $\nu_1 > \nu_3$ [188]. (For IR spectra the intensity of ν_3 is expected to be greater

Table 5. *Ranges of quoted mode frequencies* (cm^{-1}) *for isolated* H_3O^+ *and* D_3O^+ *in solids*

Mode (C_{3v} label)	H_3O^+ Peak values	Band halfwidths‡	D_3O^+ Peak values	Band halfwidths
ν_1 ⎫ ν_3 ⎭	2450–3509	80–1200†	1945–2600	350–1000†
$2\nu_2$ or $(\nu_4 + \nu_L)$	2050–2250	100	1430–1700	
ν_4	1575–1730	100–250	1160–1350	30–80
ν_2	900–1182	50–200	710–885	50–200

† Covers both modes under one band envelope.
‡ The ranges of half-widths were estimated from published spectra.

than that of ν_1 and the reverse applies to Raman spectra.) An *ab initio* calculation [329] places $\nu_3 > \nu_1$ and gives ν_3 a much greater IR intensity. Other authors have made assignments on the basis of deuteration studies: $\nu_3 > \nu_1$ [40, 73, 74] or $\nu_1 > \nu_3$ [281]. The Raman polarization behaviour of $H_3O^+SbCl_6^-$ in CH_2Cl_2 solution [46, 205] suggested $\nu_1 > \nu_3$. Calculations have also suggested that the order may even depend on the hydrogen-bond strength, with $\nu_1 > \nu_3$ for stronger hydrogen bonds [281]. Many authors judiciously refrain from making assignments.

A band in the 2050–2250 cm^{-1} region appears to be a common feature of H_3O^+ spectra and has been assigned either as $2\nu_2$ (which in some cases would require a negative anharmonicity) or $(\nu_4 + \nu_L)$ where ν_L is a librational mode. Figure 6 shows recent Raman spectra of $H_3O^+ClO_4^-$ in its room temperature phase [77]. Assignment of ν_2 is still uncertain because of overlap with ClO_4^- bands and the order of ν_1, ν_3 has not been conclusively demonstrated; otherwise this is one of the simpler spectra of H_3O^+ in a solid. The spectra of $H_3O^+SbCl_6^-$ [46, 205] (figure 7) show the least effects of hydrogen bonding, as might be expected with this anion. The modes were assigned as $\nu_1 = 3509$, $\nu_3 = 3463$, $\nu_4 = 1604$ and $\nu_2 = 1095$ cm^{-1}. The ν(O–H) modes are the narrowest observed and have the highest (reliable) frequencies, though it is of interest to note that values for H_3O^+ in the gas phase are still higher (see later).

Aside from the fundamental vibrational modes of H_3O^+, various low-frequency modes have been identified as librational (816–420 cm^{-1} range) or translational modes (< 400 cm^{-1}). The libration (R_z) about the principal axis (the C_3-axis) is likely to give the weakest intensity in both IR and Raman (if it is observed at all), but if it could be correctly identified for a particular compound, the barrier to rotation could be estimated and compared with activation energies from NMR studies. Incoherent, inelastic, neutron scattering is particularly useful for the study of librational modes since the technique

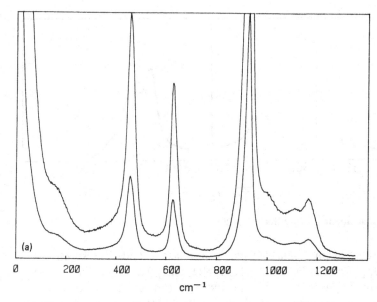

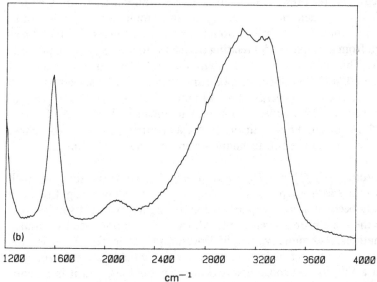

Figure 6. Raman spectra of the room temperature phase of $H_3O^+ClO_4^-$ [77] in the 50–1350 cm^{-1} region (a) and 1200–4000 cm^{-1} region (b). In (a) the upper trace is multiplied by three to enhance the detail in the 800 to 1250 cm^{-1} region.

is sensitive to large amplitude H motions. For H_3O^+ all three librational modes, R_x, R_y, and R_z, should be quite prominent. A number of studies of $H_3O^+ClO_4^-$ [70–71, 75, 76] and $H_3O^+NO_3^-$ [107] carried out between -150 °C and room temperature using inelastic and quasi-elastic neutron scattering have been published by Janik *et al*. Single, broad, librational bands were

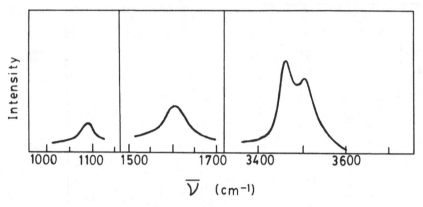

Figure 7. Raman spectra of $H_3O^+SbCl_6^-$ [46, 205].

observed at $520\ cm^{-1}$ and $560\ cm^{-1}$ in phase I and II respectively of $H_3O^+ClO_4^-$ [75] and at $847\ cm^{-1}$ in $H_3O^+NO_3^-$ [107]. The authors assumed that the peaks in the phase II perchlorate and the nitrate corresponded to R_z and did not consider the presence of the two other librational modes in the same region. The assignment may still be correct but should not be considered to be certain. A three-fold re-orientational barrier height of $19.2\ kJ\ mol^{-1}$ was estimated from a simple model relating frequency to barrier, for the phase-II H_3OClO_4. This compares favourably with the ^1H NMR activation energy of $20.1\ kJ\ mol^{-1}$ [81]. Quasi-elastic neutron scattering results were used to determine an activation energy of $6.7\ kJ\ mol^{-1}$ in phase-I H_3OClO_4 [76] and total neutron scattering cross-section measurements [71] gave $7.5\ kJ\ mol^{-1}$, values which do not agree with the NMR activation energy for this phase [81], suggesting that the two techniques were looking at different motions.

3.4.2. *Spectra of $H_5O_2^+$ and Higher Hydrates.* Clearly if the situation with respect to the spectroscopy of the isolated H_3O^+ ion is not straightforward, it can only become more complicated for the higher hydrated species. Not only are the same problems encountered, they are compounded by variable conformations and more varied hydrogen-bond strengths. Whether the vibrations of the higher hydrated proton species should be considered as a whole in a full normal coordinate treatment is dubious, and it is perhaps better to treat them as H_3O^+ (or $H_5O_2^+$ in certain cases) with attached water molecules, with all the sub-units interacting and perturbing each other.

Even discrete $H_5O_2^+$ with its variable conformations is a complicated unit. The normal modes have been worked out and illustrated for the symmetric C_{2h} *trans* configuration [96, 97], and for a C_s *gauche* configuration [49]. All these analyses assumed a centred proton in the O–H'–O bond. In these whole group models the modes can be classed as O–H stretching and bending of the two terminal H_2O units, motions of which are coordinated to meet symmetry requirements, and the stretching and bending modes of the O–H′–O unit. The positions and widths of the peripheral O–H stretching modes again depend on the hydrogen-bond strengths involved, though they are usually still above 2500 cm^{-1} and the 'H_2O' bending modes usually occur in the $1650–1750 \text{ cm}^{-1}$ range. An alternative view would be to consider the $H_5O_2^+$ as a highly distorted H_3O^+ unit attached to an H_2O unit. This may, in the general case, be better in light of the structural studies whieh show that in most cases the bridging proton, H′, is not centred and for most of the symmetrical examples the position of H′ is probably disordered.

A primary interest in $H_5O_2^+$ is centred on the spectral effects of the central strong hydrogen bond, which, when it is very short, ought to give very broad stretching mode bands [326] at much lower frequencies than for the peripheral O–H units. A brief description of the modes of an O–H–O unit have been given by Williams [34]. The properties of very strong hydrogen bonds, including their vibrational spectroscopy, have been reviewed by Emsley [30], and although he mentions $H_5O_2^+$ only briefly, he describes the analysis of its vibrational spectra as a knotty problem. The spectra must depend to some extent on the nature of the hydrogen-bond potential well, i.e. whether it is an asymmetric double minimum, a symmetric double minimum with high or low barrier, or a single broad minimum (discussed later). There seems to be no general agreement on the assignment of the O–H′–O modes although bending modes are usually assigned in the $1100–1400 \text{ cm}^{-1}$ region, ν_{sym} (O–H′–O) generally at low frequency (450 cm^{-1} region), and the very broad ν_{asym} (O–H′–O) anywhere in the $500–2000 \text{ cm}^{-1}$ region. Rozière *et al.* [32, 49, 204, 208], who have perhaps made the greatest attempts to understand the spectroscopy of $H_5O_2^+$, suggest that in some cases in the IR spectra ν_{asym} (O–H′–O) is very broad and can span the entire $2000–700 \text{ cm}^{-1}$ range. This is in keeping with the behaviour of strong hydrogen bonds in other systems; from the $\nu(\text{OH})$ *versus* $O \cdots O$ relationship [327] and a very rough relationship between halfwidth and frequency shift [326] one might expect for $O \cdots O = 245 \text{ pm}$: ν_{asym} (O–H′–O) $= 850 \text{ cm}^{-1}$, halfwidth $= 1340 \text{ cm}^{-1}$; and for $O \cdots O = 250 \text{ pm}$: ν_{asym} (O–H′–O) $= 1500 \text{ cm}^{-1}$, halfwidth $= 1310 \text{ cm}^{-1}$.

· To date there has only been one detailed single-crystal Raman polarization study of a compound established as containing $H_5O_2^+$ by neutron diffraction, i.e. $Y(C_2O_4)_2 . H_5O_2 . H_2O$ [139]. The polarization data enabled the crystal symmetry of the bands to be determined, but it was found that this still did not really aid the task of assignment. Furthermore, for this material, regions of the spectrum include the bands of a separate molecule of water and of the oxalate ion. The authors attempted to relate the root-mean-square displacements of the atoms of the $H_5O_2^+$ to the thermal ellipsoids of the neutron study, and in order to obtain reasonable agreement they chose a model which suggested that the terminal H_2O units vibrate relatively independently with the central proton vibrating parallel and perpendicular to the $O \cdots O$ direction.

Such polarization studies on single crystals (both IR and Raman) are potentially quite useful for making assignments, provided the crystal structure is known and the number of molecules in the primitive cell is not too large, though for many species this could present quite a few experimental difficulties.

The ionic conductors hydrogen uranyl phosphate (HUP) and hydrogen β-alumina, containing $H_5O_2^+$, have both been studied extensively by vibrational spectroscopy. HUP has been studied by IR and Raman spectroscopy [154, 155] as a function of temperature and by the complementary technique of inelastic neutron scattering (INS) [153], which observes principally motions of hydrogen-containing species, which in this material consist of $H_5O_2^+$, $H_3O^+ . H_2O$ and H_2O. Obviously this system is very complex, but the authors have attempted assignments. It is clear that INS would be a very useful technique for further studies and might help to settle some of the assignment problems of $H_5O_2^+$, but it would be advisable to look first at well characterized materials containing only $H_5O_2^+$. Hydrogen β-alumina has also been studied by IR and Raman [221, 222, 224, 226]; depending on the degree of hydration the species present are thought to be H_3O^+, $H_3O^+ . H_2O$, $H_5O_2^+$ and even $H_7O_3^+$. The authors claim to be able to distinguish H_3O^+ and $H_5O_2^+$ in samples containing both types of ion. On the other hand, one IR paper even suggests that there are no hydrated proton species present in hydrogen β-alumina [225].

Much weight has been placed on the interpretation of IR spectra in support of the presence of hydrated proton species in minerals [247, 248, 251, 252, 254–256] and zeolites [265, 266]. However, these results alone do not provide conclusive proof. There may be cases where the spectra do strongly suggest hydrated proton species (for instance the appearance of several new bands, besides those due to adsorbed H_2O, in a hydrated zeolite known to contain acidic groups when compared to one which does not have acidic groups [266]), but assignment to particular forms can only be speculative.

Overall one cannot help but get the impression that a great many of the claimed assignments made for $H_5O_2^+$ should be labelled 'tentative only'. Gillard and Wilkinson [172] introduced a general way of categorizing the $H_5O_2^+$ spectra by quoting frequency ranges for four bands which are apparent in most of the spectra. If one does this one finds that there is in fact not much difference between these ranges and those of H_3O^+. Again one is struck by the different frequencies given for the same materials by different authors. Williams [34] went on to categorize the spectra of $H_7O_3^+$ and $H_9O_4^+$ into four frequency bands in much the same way. All this is still of limited value, but it does serve to emphasize that really one cannot establish a set of rules for distinguishing the spectra of the different species. Though spectra of several materials containing $H_7O_3^+$ and $H_9O_4^+$ clusters have been obtained [44, 45, 56, 100, 207, 211, 274] again the same complexity holds. More frequently, however, in such cases the authors have sought an interpretation in terms of H_3O^+ and H_2O sub-units.

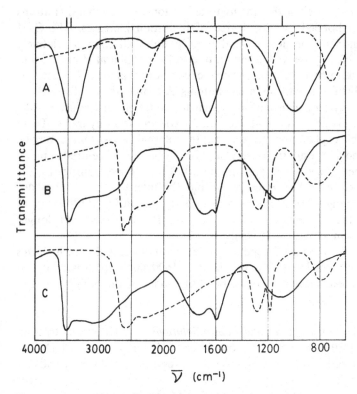

Figure 8. Infrared spectra of (A) HSbCl$_6$.2H$_2$O, (B) HSbCl$_6$.3H$_2$O and (C) HSbCl$_6$.4H$_2$O, redrawn for purposes of comparison from the figures of Ortwein and Schmidt [207]. The broken curves are for the deuterium-substituted species. The positions of the Raman bands of H$_3$O$^+$SbCl$_6^-$ [46, 205] are indicated at the top of the figure.

Frankly the authors of this review remain very sceptical about the merit of attempting detailed assignments of the higher proton hydrates. In fact the effects of increasing hydration are perhaps best seen in comparisons of the spectra of series of hydrates of the same anion, which may be done for example with SbCl$_6^-$ [32, 46, 205, 207], PtCl$_6^{2-}$ [190], Cl$^-$ [44, 45, 47] and ClO$_4^-$ [32, 78, 100]. The best example to consider is the series of hydrates of SbCl$_6^-$ (see figure 8). One would expect little or no hydrogen bonding to the anion in this series; thus complications from this source are removed and one need only be concerned with the effects of hydrogen bonds within the hydrated proton clusters. In fact, as mentioned earlier, the H$_3$O$^+$ salt has the narrowest of all O–H stretching bands [46, 205] with quite high frequencies. The IR spectra of the higher hydrates [32, 207] immediately show broader bands, but one can see two effects:

(1) Comparison of the known structures of the di- and tri-hydrates [206, 209] and possible structures for the tetra-hydrate shows that there has to be an increase in the ratio of internal H atoms (i.e. H atoms involved in

hydrogen bonds entirely *within* the hydrated cations) to peripheral H atoms (non-hydrogen bonded) as the degree of hydration increases. The O–H stretching region of the di-, tri- and tetra-hydrates all show a relatively narrow component close to 3500 cm^{-1}, due to the peripheral H atoms, and in the tri- and tetra-hydrates a much broader component at lower frequencies, which can only be due to the internal H atoms. Although the O–H stretching band for the dihydrate is broadened somewhat to the lower frequency side, the broadening is not as dramatic as for the tri- and tetra-hydrates. There are two possible explanations for this: (a) the internal O–H–O bond may be very strong and as a result the ν(OH) band is very broad and centred at much lower frequencies, possibly underlying the whole of the spectrum; or (b) the species behaves more like $H_3O^+ . H_2O$ and the internal O–H$\cdots$O is not particularly strong. (The X-ray structure [206] gave an O$\cdots$O distance of 252 pm (note the refinement was not very good, $R = 0.116$) which is longer than the criterion allows for $H_5O_2^+$).

(2) In the 1600–1700 cm^{-1} bending mode region the tri-and tetra-hydrates again show an extra band at approximately 1600 cm^{-1} which is relatively sharp and is presumably due to the bending modes of the peripheral H_2O units.

3.4.3. *Matrix-Isolated Species and Glassy Solutions.* The IR studies by Ault and Pimentel [330] of isolated 1:1 complexes of HCl/H_2O and DCl/D_2O in solid nitrogen matrices at 25 K showed the presence of $H_2O . HCl$ and not the ion pair $H_3O^+Cl^-$. Later studies showed that the same was true for $H_2O . HCl$ in argon matrices [331], for $H_2O . HBr$ in nitrogen and argon matrices [332] and for $H_2O . HF$ complexes [333]. Similarly, Ritzhaupt's and Devlin's [334] IR studies of HNO_3/H_2O in solid argon matrices at 10 K showed that in isolation the preferred species is $H_2O . HNO_3$ and that $H_3O^+NO_3^-$ ion pairs were only stable when solvated by other H_2O molecules. All this was inferred from observation of the concentration-dependent changes of the HNO_3 and NO_3^- bands; bands of H_3O^+ itself were not observed. These results are interesting, from the point of view of solids, in that they show that the ionic, crystalline monohydrates of these acids must be stabilized by the crystalline environment and a gain in lattice energy.

Kanno and Hiraishi [335, 336] have recently studied glasses, consisting of aqueous solutions of HCl, HBr and HI, by Raman spectroscopy. Although these are solids at the temperatures studied, they are best regarded as frozen pictures of the liquid state. Bands at about 1230 and 2800 cm^{-1}, comparable with liquid solution frequencies, were assigned to ν_2 and ν_1 modes of H_3O^+. These results and the matrix isolation studies will also be of relevance to the later discussion of the nature of the hydrated proton in solution.

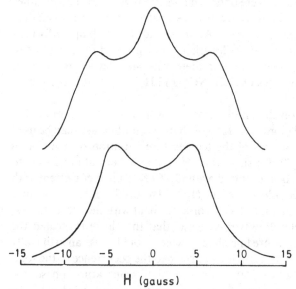

H (gauss)

Figure 9. Calculated ^{1}H NMR line-shapes for (top) a rigid, three-spin, H_3O^+ group [16] and (bottom) a rigid, 2-spin, H_2O group [358], taking into account the broadening effects of other neighbouring ^{1}H atoms in the lattice.

3.5. NMR Studies

NMR studies of the hydrated proton in solids can give useful structural information but they are also of particular interest since NMR is one of the few techniques which can yield information about the molecular dynamics. Unfortunately there have not yet been many reports dealing with this area and certainly much more useful work could be done.

3.5.1. *^{1}H NMR Line-shapes and Second Moments.* ^{1}H NMR studies in fact provided one of the earliest proofs of the existence of H_3O^+. ^{1}H NMR line-shapes in solids arise from dipole–dipole interactions of the spin $\frac{1}{2}$ ^{1}H nuclei, and the magnitude of this interaction is inversely proportional to the cube of the interproton distance; consequently H_2O and H_3O^+, a pair and a triangle of spins respectively, give quite different line-shapes when they are not in motion (see figure 9). Kakiuchi *et al.* [17, 79, 85] obtained line-shapes for $HClO_4 . H_2O$ below 140 K which could only be consistent with the presence of a triangle of protons as in H_3O^+ and not $H_2O . HClO_4$. At the same time Richards and Smith [16] observed the $H_3O^+ClO_4^-$ line-shape at 90 K and also line-shapes at 90 K consistent with H_3O^+ in several other materials [16, 118]: $H_3O^+NO_3^-$, $H_3O^+HSO_4^-$, $(H_3O^+)_2PtCl_6^{2-}$, $H_3O^+HSeO_4^-$ and $(H_3O^+)_2SO_4^{2-}$. (The result for the last compound was a little ambiguous but the X-ray structure later confirmed this formulation [117].) Rigid-lattice

174   C. I. Ratcliffe and D. E. Irish

second moments of 30–32 G^2 were obtained consistent with calculated values for H_3O^+, and the $H \cdots H$ distance was estimated to be approximately 172(2) pm from these early studies. Andrew and Finch [86] performed calculations of line-shapes for isosceles triangles of protons for comparison with these results and showed that if there was any deviation from an equilateral triangle in the cases of $H_3O^+NO_3^-$ and $H_3O^+ClO_4^-$ it could not be more than 10 per cent.

The effect of motion on the dipole–dipole tensors is to cause a reduction in line-width and second moment (M_2), which in favourable cases can be used to identify the motion. Studies of the temperature variation of line-shapes and M_2 of $H_3O^+ClO_4^-$ [79–83] show that M_2 reduces from about 32 G^2 to about 10 G^2, the change being centred around 150 K. This is consistent with a rotation about the three-fold axis of the H_3O^+. The results of Herzog-Cance et al. [83] give M_2 values of about 6 G^2 in disagreement with the other studies; inspection of their results, however, suggests that they have truncated the wings of the higher temperature line-shapes, which would cause an artificially low value for M_2. These NMR studies also showed the occurrence of the solid I–II phase change close to 250 K. In the high-temperature phase the line-shapes and M_2 suggest a pseudo-isotropic re-orientation of H_3O^+.

Results for the $PtCl_6^{2-}$ salt [83] also suggest a three-fold re-orientation with the M_2 reduction centred around 140 K, indicating a slightly lower activation energy than for $H_3O^+ClO_4^-$. A high-temperature phase change for the $PtCl_6^{2-}$ salt at approximately 367 K was also detected. H_3O^+ apparently remains rigid up to the respective melting points in the NO_3^- [83, 116], HSO_4^- and SO_4^{2-} [116] salts. However, a second study of $H_3O^+HSO_4^-$ suggests that some motion begins just before melting [87]. Line-shapes and M_2 results for two materials formulated as $(H_3O)X(SO_4)_2 \cdot H_2O$ where $X = In$ or Tl are not inconsistent with the presence of H_3O^+ [144], and the M_2 values indicate that a rapid general re-orientation occurs above approximately 200 K.

Some estimates of the activation energies (E_a) for the three-fold motion were also obtained from these temperature variation studies: for $H_3O^+ClO_4^-$, 12.6 [80], 19.7 [82] and 19.0 [83] kJ mol^{-1}, and for $(H_3O^+)_2PtCl_6^{2-}$ 11.4 kJ mol^{-1} [83]. Studies of the unusual systems $(H_3O^+)_2(NO_2^+)_9(ClO_4^-)_{11}$ and $(H_3O^+)(NO_2^+)(ClO_4^-)_2$ [83] showed that H_3O^+ in these systems, is not rigid even at 90 K. Note that the narrowed line-shapes do not prove the presence of H_3O^+, though the formulation of $(H_3O^+)(NO_2^+)(ClO_4^-)_2$ was recently shown to be correct in an X-ray study [359].

Low-temperature line-shape studies of materials of formula $H(H_2O)_n$ SbO_3 ($n = 1, 0.92, 0.2$), which are proton conductors, have confirmed the presence of H_3O^+ [213]. The hydrated uranate of empirical formula $UO_3 \cdot 2H_2O$ has a low-temperature line-shape which corresponds best with a formula $H_3O^+ \cdot UO_2^{2+} \cdot O^{2-} \cdot OH^-$ [263]. The line-shapes of sodium meta-silicate nona-hydrate were very tentatively interpreted in terms of $Na_2(H_3O)_2SiO_4 \cdot 6H_2O$ [178]. This case illustrates the limitations of the line-shape method when the system is not simple: i.e. overlapping signals from protons in different types of molecule, and the possibility of the same overall line-shape resulting from combinations of different constituents. Another example is $(H_3O)Ga_3(OH)_6(SO_4)_2$ for which M_2 studies [147] showed a line narrowing centred at approximately 80 K. While these results show that some motion was occurring at low temperature, they do not permit a determination of the exact nature of the motion, nor do they prove convincingly the presence of H_3O^+. ^1H NMR studies have also been used in studies of minerals in attempts to show the presence of H_3O^+. Again, however, the results have generally been inconclusive, because the H_3O^+ protons (assuming them to be present) are never completely isolated from other protons in H_2O or OH^- groups [248], and the H_3O^+ may not have been rigid at the temperature of the study.

While ^1H line-shapes may be sensitive to different configurations of nuclei in the rigid case for simple molecules, they are generally notoriously insensitive to fine details of motion, except in simple cases, and particularly so when one motion is already very fast (e.g. different motions can give similar narrrowed spectra). It should also be emphasized that activation energies, E_a, obtained from line-shape/M_2 studies are considered to be much less reliable than those obtained from spin–lattice relaxation results. Herzog-Cance et al. have obtained ^1H line-shapes and M_2 values between 90 K and room temperature for $H_5O_2^+$ in Br^-, ClO_4^- and $SbCl_6^-$ compounds [54] and for $H_7O_3^+$ in NO_3^-, ClO_4^-, $Cl_2C_6H_3SO_3^-$ and $C_6H_5SO_3^-$ compounds [101]. These results are interesting and useful (only two other ^1H NMR spectra of $H_5O_2^+$ have been obtained; $H_5O_2ClO_4$ at 90 K [118] and possibly $(H_5O_2)_2SO_4$ at 77 K [87]), but the authors appear to have interpreted the results far beyond the limits of the technique. What they do show is that H_5O_2Br is rigid below 100 K, $H_5O_2ClO_4$ is rigid below 160 K and above these temperatures both show evidence of some kind of motion before they melt. In both cases the slow reduction in M_2 with temperature is not typical of a single activated process. Their results for $H_5O_2SbCl_6$ show that there is already considerable motional narrowing even at 90 K, and two other motions are activated at higher temperature (M_2 changes are centred at 170 and 270 K). The models which were proposed for these motions may be plausible, but bearing in mind the limitations of the technique, they should be regarded as very tentative. $H_7O_3NO_3$ and $H_7O_3ClO_4$ appear to be rigid below 130 K, $H_7O_3Cl_2C_6H_3SO_3$ and $H_7O_3C_6H_5SO_3$ at 90 K or slightly lower, and all show motional narrowing before melting, but again a unique analysis is not possible. (It is of interest to see that the line-shapes shown for 90 K for these four $H_7O_3^+$ compounds have some very distinct differences, presumably related to the very variable geometry of this ion.)

^1H line-shapes/M_2 studies of the clathrate structure of H_3O . PF_6 . HF . $4H_2O$ [236] showed that above 175 K there was rapid diffusion of the protons. This process appears to have been enhanced by the presence of the excess acid

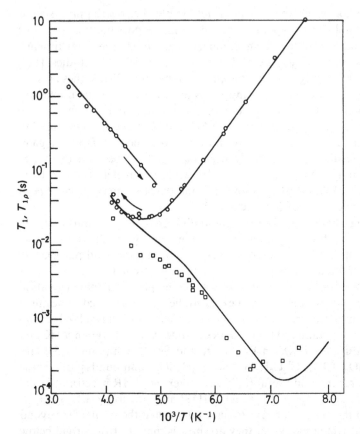

Figure 10. Logarithmic plots of ^{1}H T_1 (circles) and $T_{1\rho}$ (squares) *versus* reciprocal temperature for $H_3O^+ClO_4^-$ [81]. The arrows indicate cooling or heating in the region of hysteresis. The full lines indicate fits to the T_1 results and a calculation of the expected $T_{1\rho}$ behaviour based on the T_1 curve parameters.

protons in the cage structure, since diffusional narrowing in ice-Ih and other gas clathrate hydrates usually occurs at somewhat higher temperatures [337].

3.5.2. *Relaxation Studies.* Studies of ^{1}H spin–lattice relaxation times (T_1) *versus* temperature are very useful in determining activation energies (E_a) for motions in solids, though they may be less helpful than line-shapes in determining the nature of the motion. T_1 measures the rate of recovery of the net magnetization along the magnetic field direction to its equilibrium state, after the system has been perturbed by a resonant radiofrequency pulse. T_1 is related (by the theory of Bloembergen *et al.* [338] as extended by others [339, 340]) to the correlation time (τ_c) of the re-orientation process. τ_c is a function of temperature (T) and the activation energy (E_a)

$$(T_1)^{-1} \propto f(\omega_0 \tau_c)$$
$$\tau_c = \tau_c{}^0 \exp(E_a/RT)$$

where ω_0 is the proton Larmor frequency. Plots of $\ln(T_1)$ *versus* $1/T$ are then linear on either side of the T_1 minimum (where $\omega_0 \tau_c$ is either very large or very small compared with one) and the slopes are simply related to E_a. To date only a few hydrated proton species have been studied by this technique.

T_1 results for $H_3OGa_3(OH)_6(SO_4)_2$ [147] show a minimum at 135 K, but the slopes on either side give quite different values for E_a; 3.35 and 10.0 kJ mol^{-1} on the low and high temperature sides respectively. This is not typical of a single activated process. There is a second, shallow minimum at 300 K which was ascribed to translational diffusion in the lattice, with an E_a of 5.86 kJ mol^{-1}. Further studies, which included $T_{1\rho}$ and T_{1D} relaxation parameters [422] later gave these three E_a values as 2.09, 7.53 and 9.20 kJ mol^{-1} respectively [148]. The authors assigned these to specific motional processes, but in view of the peculiarities involved and considering all the data on this material, the results do not really justify saying much more than that the low temperature T_1 minimum probably involves an as yet unspecified motion of H_3O^+. (^{2}H NMR results on the related material oxonium alunite are discussed later).

The ^{1}H T_1 studies by O'Reilly *et al.* [81] on H_3OClO_4 (figure 10), in conjunction with ^{2}H results, have shown that in the low-temperature phase the H_3O^+ ion performs three-fold re-orientation with $E_a = 20.2$ kJ mol^{-1}. In the high-temperature phase the motion is pseudo-isotropic with an $E_a = 17.5$ kJ mol^{-1} (^{2}H T_1 results gave $E_a = 15(2)$ kJ mol^{-1}). Here the evidence is sufficient to be quite specific. Also of significance was the observation of hysteresis effects in ^{1}H T_1 in the region of the phase transition.

^{1}H T_1 studies of $CH_3C_6H_4SO_3^-H_3O^+$ [282] gave an E_a of 51.6 kJ mol^{-1} assigned to the three-fold re-orientation of the H_3O^+ ion. Although the motional assignment here was only inferred from a number of strong arguments, ^{2}H line-shapes have since shown this to be correct (see later).

The fast-proton conductors hydrogen uranyl phosphate (HUP) and arsenate (HUAs) have also been studied by proton T_1 [156]. The results show a phase transition in both materials, at 274 K in HUP and 302 K in HUAs. In their low-temperature phase two motions were postulated: a fast one with $E_a = 28$ kJ mol^{-1}, probably three-fold re-orientation of H_3O^+ and a slow one, probably self-diffusion with an E_a which increases from 30 kJ mol^{-1} at 220 K to 60 kJ mol^{-1} at the phase transition. In their high-temperature phase both compounds showed a motion thought to be rapid translational diffusion of hydrogens with an $E_a = 20$ kJ mol^{-1}. The authors proposed a mechanism for the proton translation involving a sequence of proton hops between adjacent molecules and the necessary re-orientations to allow the process to continue.

$H_5O_2^+AuCl_4^- \cdot 2H_2O$ has been extensively studied by ^{1}H and ^{2}H NMR and ^{35}Cl NQR [200]. These results show that phase changes occur at 290 K (solids I–II) and 218 K (solids II–III) for the ordinary material. In the deuterated material the II–III transition occurs at 252 K and the II–I transition was not detected below 300 K. The ^{1}H T_1 results were interpreted in terms of a jumping motion of the bridging proton in the $H_5O_2^+$ with an $E_a = 49.0$ kJ mol^{-1} in phase III and $E_a = 23.8$ kJ mol^{-1} in phase II. (The neutron structure shows phase I to have an $H_5O_2^+$ which is better described as $H_3O^+ . H_2O$, with the bridging proton in disordered positions on either of the two oxygens [199].) Rotating frame relaxation data $(T_{1\rho})$ indicated two motions: one assigned to the same jumping process as above, and the other to a proton-exchange process requiring a three-fold re-orientation of the H_3O^+ unit and translational diffusion of H_2O molecules between sites having orientations which differ by 90°. This second assignment was suggested to accommodate the ^{2}H NMR results which showed for phase II at 298 K that all the deuterons in the unit cell were equivalent. This could only be achieved by rapid exchange. Considering the complexity of this system, its very hygroscopic nature and the fact that the structures of the low-temperature phases are not known, these assignments should be considered as tentative.

Recent ^{2}H T_1 studies of $D_3O^+CF_3SO_3^-$ have indicated a single motion with an E_a of 28.5 kJ mol^{-1} [88]. ^{2}H line-shapes showed that the motion was three-fold re-orientation.

A striking feature of the values for E_a obtained for three-fold H_3O^+ (or D_3O^+) motion is their great variability; this reflects the different hydrogen-bond strengths and the symmetry of the environment of the ion.

3.5.3. ^{2}H NMR Line-shape Studies.

In ^{2}H NMR [341] the line-shape depends on the quadrupole coupling tensor of the ^{2}H nucleus which is determined by the electric field gradient at the nucleus. In a static situation this usually means that the largest (or zz) component of the tensor lies along or close to the bond in which the ^{2}H atom is involved. The principal components of the tensor are characterized by the quadrupole coupling constant $(QCC) = e^2qQ/h$ and the asymmetry parameter η and in a powder spectrum these give rise to three or sometimes two (if $\eta = 0$) pairs of discontinuities separated by:

$$\nu_{zz} = 3(QCC)/2$$
$$\nu_{yy} = 3(QCC)(1+\eta)/4$$
$$\nu_{xx} = 3(QCC)(1-\eta)/4.$$

Also the quadrupole coupling tensor can be quite sensitive to the strength of any hydrogen bond in which the ^{2}H might be involved. A general empirical relationship between the QCC and the $D \cdots O$ distances in $O-D \cdots O$ hydrogen bonds is well established [342]:

$$e^2qQ/h = 317.1 - 3.0 \times 200.2/R^3{}_{D-O} \text{ kHz} \qquad (R \text{ in } \mathring{A}).$$

Since the line-shape derives from single nuclei rather than from interactions between nuclei (as with ^{1}H) ^{2}H may be less useful for discriminating different species in static situations, but it is generally far more sensitive for characterizing motions. Rapid re-orientations produce a narrowed line-shape whose

QCC and η correspond to an average of the quadrupole tensors at each site sampled by the motion [341]. Particular motions such as three-fold re-orientation of D_3O^+ or two-fold re-orientation of D_2O produce very distinct line-shapes. One can also derive geometric information about the angle between the rotation axis and the O–D bond from this change in line-width. Very little work has been done so far using this technique, but it is a very promising area, especially since the advent of the FT NMR methods.

O'Reilly *et al.* [81], using continuous wave techniques, showed conclusively (in agreement with ^1H line-shapes/M_2 studies) that in D_3OClO_4 a three-fold re-orientation occurs in the low-temperature phase; the QCC reduced from 212(5) kHz (rigid case) to 72(2) kHz with $\eta = 0$. A single sharp line due to rapid general re-orientation was seen in the high-temperature phase. More recent studies using FT NMR have revised the rigid value to 165.6 kHz with $\eta = 0.13$ [88].

In $D_5O_2AuCl_4 \cdot 2D_2O$ phase II, a single crystal ^2H NMR spectrum at 298 K showed only one type of ^2H (with $QCC = 27$ kHz and $\eta = 0.20$) indicating fast exchange between all D positions in the lattice [200]. (A powder line-shape at the same temperature gave $QCC = 26$ kHz and $\eta = 0.26$). The complex motions postulated to bring about this averaging were mentioned earlier. Clearly this system is not simple and is complicated because of the different types of ^2H (i.e. distinct H_2O and $H_3O^+ \cdot H_2O$ [199]). The line-shape began to broaden below 270 K indicating the freezing in of some motion.

A similar situation where all the deuterons become equivalent because of rapid exchange and diffusion occurs in the fast-proton conductors DUP and DUAs mentioned earlier in the proton studies [157]: Single axially-symmetric ($\eta = 0$) line-shapes with $QCC = 39.2$ and 38.2 kHz respectively were observed for the two materials in their high-temperature phase. The same line-shapes were observed in the low-temperature phase close to the transition but disappeared at lower temperatures presumably due to the slowing down of the motions. (The phase changes in these deuterated materials were at 260 K (DUP) and 290 K (DUAs).) A series of concerted motions involving transfer of ^2H across hydrogen bonds and re-orientation of D_3O^+ and D_2O units was postulated.

The QCCs and ηs for the individual deuterons in D_3ODSO_4 at 77 K have been determined [115] by the different technique of $^1H/^2H$ nuclear quadrupole double resonance. The oxonium ion is rigid (recall earlier ^1H line-shapes/M_2) and the interest here is in the differences in the parameters for each ^2H, none of which are equivalent (see table 6). (It was also found that there are small variations when the group is HD_2O^+ or H_2DO^+.) The numbers demonstrate the great sensitivity of the QCC to the hydrogen-bond strength. The results obtained by this particular method also circumvent a problem which is sometimes encountered with ^2H NMR powder line-shapes (which are more commonly obtainable) where line-shapes for all three sets of parameters (and for the DSO_4^-) would be overlapping.

Much more recently ^2H FT NMR powder line-shapes have been obtained

Table 6. ^2H NMR quadrupole coupling constants and asymmetry parameters for D_3O^+ in solids

	Static		C_3 rotation	
	QCC (kHz)	η	QCC (kHz)	η
$D_3O^+DSO_4^-$ [115]	140.5	0.154	—	—
	166.0	0.125	—	—
	177.2	0.153	—	—
$D_3O^+ClO_4^-$ [81]	212 (± 5)	?	72 (± 2)	0
$D_3O^+ClO_4^-$ [88]	165.6	0.130	73.98	0
$D_3O^+CF_3SO_3^-$ [88]	≈ 154†	?	74.79	0.126
$D_3O^+ClO_4^-$. 18-crown-6 [88]	141.0	0.111	78.21	0.104
$D_3O^+C_6H_4CH_3SO_3^-$ [88]	147.3	0.158	61.33	0.141

† A superimposition of line-shapes from the three different ^2H nuclei – approximate value.

for D_3OClO_4 (figure 11) (the earlier continuous-wave rigid-lattice value was much too large), D_3OClO_4 . 18-crown-6 complex, $D_3OCF_3SO_3$ and $D_3OCH_3C_6H_4SO_3$ [88]. Parameters are collected in table 6. In all the low-temperature spectra the discontinuities are less distinct than would be expected for a single type of ^2H. This is especially so for $D_3OCF_3SO_3$ where only a rough estimate of the average QCC could be made. It is of interest to note that in some cases the three-fold-re-orientation-averaged line-shape has a non-zero η. This also is a consequence of the inequivalent environments of the three positions sampled by the ^2H in its reorientation [341]. (If all sites were equivalent η would be zero.)

A recent ^2H study of oxonium alunite [249], nominally $D_3OAl_3(OH)_6(SO_4)_2$ for 100 per cent substitution of K^+ alunite by D_3O^+, as a function of temperature, showed two superimposed lines with different shapes at room temperature: a broad one which can be assigned to the OD groups and a narrower one, without structure, which broadened to an apparently rigid line-shape at 77 K. Dilute ^1H 'magic-angle' spinning NMR results (discussed below) indicated the presence of D_3O^+ and D_2O, and it seems that the narrower ^2H line-shape is due to both these molecules undergoing some kind of general re-orientation. No evidence for three-fold re-orientation of D_3O^+ was seen.

Another very recent ^2H study of $D_5O_2^+CF_3SO_3^-$ [88] is of even more interest since for the first time this gives some concrete information on an $H_5O_2^+$ containing a very strong hydrogen bond. The rigid case (see figure 12) at low temperature shows two superimposed patterns; the broader outer one is very similar to that of hydrogen-bonded water and corresponds to the four peripheral deuterons, whereas the much narrower central line-shape

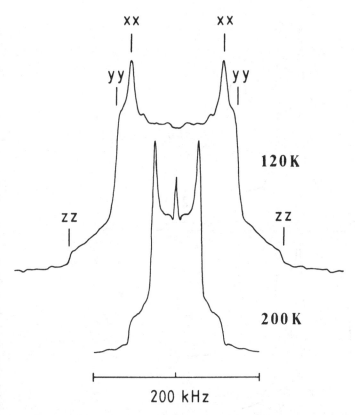

Figure 11. ^{2}H NMR line-shapes of the low-temperature phase of polycrystalline $D_3O^+ClO_4^-$ for the rigid (120 K) and fast C_3 re-orientation (200 K) cases [88]. The approximate positions of the pairs of discontinuities are indicated for the rigid case.

(which has an approximately 1:4 intensity ratio with the broad line as expected) corresponds to the strongly hydrogen-bonded central deuteron. For the broad line $QCC = 200.8$ kHz, $\eta = 0.092$ and for the narrow line $QCC = 39.39$ kHz, $\eta = 0$. Perhaps the most interesting feature, however, is the nature of the first motion which begins as the temperature is increased. At 240 K the line-shape can be readily explained as a superposition of the same narrow line-shape (central deuteron) and a motionally averaged line-shape for the peripheral deuterons which corresponds very well with two-fold flips of the two end D_2O groups. Assuming this model the H–O–H angle between the peripheral deuterons is calculated to be about 110.5°. Clearly the $D_5O_2^+$ is not behaving like $D_3O^+ . D_2O$. Not until even higher temperatures does the narrow component being to show signs of motional narrowing, but unfortunately at present one can only speculate about the motion involved.

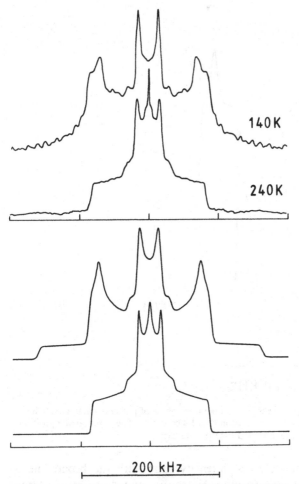

200 kHz

Figure 12. ^{2}H NMR line-shapes of polycrystalline $D_5O_2^+CF_3SO_3^-$ [88] for the rigid case (140 K) and showing the effects of two-fold flips of the two end 'D_2O' units (240 K). The narrower central line-shape in the 140 K spectrum is still present at 240 K and is due to the central $^2H'$. The lower half of the figure shows simulated line-shapes based on a 4:1 intensity ratio for the component line-shapes of the peripheral and central ^{2}H nuclei.

3.5.4. *^{1}H Chemical Shifts.* ^{1}H NMR measurements of solutions yield very narrow lines corresponding to the isotropic chemical shifts of the nuclei being studied; these are very characteristic of the molecule in which they are located. Only more recently have techniques been developed which permit ^{1}H isotropic spectra to be obtained in solids. Dilute ^{1}H 'magic-angle' spinning NMR has been used to obtain such spectra for some H_3O^+ and $H_5O_2^+$ species [84]; H_3OClO_4 (spinning not necessary for the room temperature phase), H_3O^+ alunite (mentioned earlier), $(H_5O_2)_3PW_{12}O_{40}$ and $H_5O_2Y(C_2O_4)_2 \cdot H_2O$.

Table 7. *^1H NMR chemical shifts in solids from dilute ^1H MAS experiments* [84]

	δ (ppm, TMS = 0)	
$H_3O^+ClO_4^-$ (non-spinning)	10.7	
H_3O^+ replacing K^+ in $KAl_3(OH)_6(SO_4)_2$	11.4	
$(H_5O_2)_3(PW_{12}O_{40})$	8.6	⎫ Assigned to peripheral
$(H_5O_2)Y(C_2O_4)_2 \cdot H_2O$	6.5	⎬ protons
	10.1	⎭
H_2O in crystals	4.3–7.6	
H_3O^+ in non-aqueous solution [240, 241, 343]	9–12	

Shifts are given in table 7. More oxonium solids must be studied with this technique, but it seems quite certain from these results and *ab initio* calculations that the H_3O^+ chemical shifts fall into a range at a lower field than the range for H_2O. The range for $H_5O_2^+$ peripheral proton shifts overlaps the H_3O^+ range and the low-field end of the H_2O range.

3.6. *Proton Conduction*

Proton conduction has been found to occur in several stoichiometric compounds of the hydrated proton; $HUO_2PO_4 \cdot 4H_2O$ (HUP) [159–162], HUAs [167, 168], H_3OClO_4 [94], and others which have a more variable composition, though their hydrated proton content is still high: H-β-alumina [229]; H-β'' alumina [228]; $HSbO_3 \cdot xH_2O$ [212, 214, 215]; $HSbTeO_6 \cdot xH_2O$ [216]. (Typical empirical formulae for H-β- and β'' alumina are $[1.24H_2O, 11Al_2O_3, 2.6\ H_2O]$ and $[0.84\ H_2O, 0.84\ MgO, 5\ Al_2O_3, 2.8\ H_2O]$ respectively [228]. The protons of the nominally H_2O units at the front of each formula are the acidic protons.) There are also a number of solids in which the hydrated proton, even in minute quantities, may take part in the conduction mechanism; examples include ice, clathrate hydrates and other materials where there is an extended hydrogen-bonded water network [344]. The general topic of proton conduction in solids was reviewed thoroughly by Glasser [344] in 1975, but at that time this did not include any cases involving H_3O^+ other than ice.

In more recent years attention has been focused on the fast proton conductors, where 'fast' generally means a conductivity greater than about $10^{-5}\ \Omega^{-1}\ cm^{-1}$. These include the high-temperature phases of the three solids HUP [159–162], HUAs [168], $6 \times 10^{-3}\ \Omega^{-1}\ cm^{-1}$ at 298 K and 310 K respectively, and H_3OClO_4 [94] $3 \times 10^{-4}\ \Omega^{-1}\ cm^{-1}$ at 298 K, also H-β''-alumina [228], $10^{-2}\ \Omega^{-1}\ cm^{-1}$ at 298 K, $HSbO_3 \cdot xH_2O$ [214], $3.2 \times 10^{-3}\ \Omega^{-1}\ cm^{-1}$ at 298 K, and $HSbTeO_6 \cdot xH_2O$ [216], $10^{-2}\ \Omega^{-1}\ cm^{-1}$ at 305 K. A main interest in fast proton conductors is centred around their potential for use as

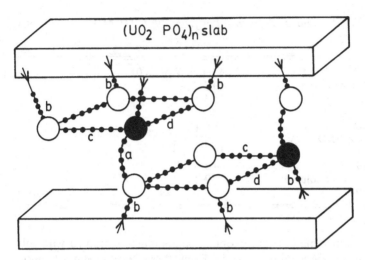

Figure 13. Schematic figure [154] of the layered structure of $H_3OUO_2PO_4 \cdot 3H_2O$ (HUP), showing the oxygen atoms of H_2O (open circles) and H_3O^+ ions (full circles) which make up the proton-conducting layer, and the potential hydrogen-bond linkages (large dots). There is a statistical distribution of H atoms among the more numerous hydrogen-bond sites. (The hydrogen bonds (b) are to PO_4 oxygen atoms.)

high-temperature fuel-cell solid electrolytes [228]. Structural [150, 151, 164–166, 218, 219], NMR [156, 157, 213], vibrational spectroscopy [153–155, 221, 222, 224–226], and quasi-elastic neutron scattering (QNS) [152, 223] techniques have all been used in attempts to understand the conductivity mechanism in such materials.

In the high-temperature phases of HUP and HUAs the conduction occurs principally in the two-dimensional layers of interconnecting H_3O^+ and H_2O. One feature of the layer (figure 13), which is of significance to the conduction mechanism, is that there are only nine excess acid protons disordered over ten hydrogen-bond sites. Conductivity measurements on single crystals of HUP [161] showed that conduction perpendicular to these layers was reduced by a factor of at least 100. In both materials their low-temperature phases have reduced conductivities. Similarly in the low-temperature phase of $H_3O^+ClO_4^-$ the conductivity is reduced by a factor of 100 or more [94]. In H-β''-alumina also the conduction occurs in two-dimensional layers, and it is of significance that this material has more water molecules per proton than ordinary H-β-alumina, which has a much lower conductivity [228], $10^{-11} \Omega^{-1}$ cm^{-1} at 298 K. As mentioned earlier, the NMR activation energies of 20 (1) kJ mol^{-1} for rapid H diffusion found for both HUP and HUAs (high-temperature phases) [156] are different from that obtained from the conductivity studies [159] of 31 (1) kJ mol^{-1}, suggesting that different processes are responsible for the observed results. This must be connected with the requirement for macroscopic proton transport in the conduction

The Nature of the Hydrated Proton

mechanism. Quasi-elastic neutron scattering (QNS) results for HUP (high-temperature phase) suggest rapid re-orientation of H_3O^+ with a low E_a at 307 K and slower re-orientation of H_2O with $E_a = 20$ (5) kJ mol^{-1} [152]. A third even slower motion was thought to be an anisotropic translational motion consistent with a proton jump distance of approximately 200 pm.

Three principal models have been developed to explain proton conduction in solids:

(1) A simple hopping of H_3O^+ from one site to a vacant site, each hop creating a vacancy for another H_3O^+. This is most likely when there is not an extended hydrogen-bonded network of H_2O and H_3O^+ molecules.

(2) A mechanism similar to the Grotthuss mechanism proposed for acid solutions (see [345, 346]), involving a succession of proton transfer (tunnelling) steps between adjacent hydrogen-bonded O atoms, together with a rotation of H_2O molecules so that the process can continue. (The latter re-orientation is the rate-determining step.) This mechanism requires disorder in the hydrogen-bonded network.

(3) A 'vehicle mechanism' involving translational diffusion of whole 'loaded vehicle' H_3O^+ ions and a counter diffusion of 'empty vehicle' H_2O molecules [167, 347].

The second mechanism is still perhaps the most favoured, particularly for the fast-proton conductors (e.g. in HUP [163]). The third mechanism is gaining more recognition, however, particularly since it allows for motion of the O atoms; ^{18}O tracer diffusion experiments on HUAs have shown that the H_3O^+ and H_2O oxygens in this material completely equilibrate with an aqueous solution [167]. Whichever model one chooses to favour, the H_3O^+ ion is central to the conduction mechanism.

In the case of the fast-proton conductivity of H_3OClO_4 [94] a model was suggested in which the proton is transferred from H_3O^+ to an oxygen of ClO_4^-, followed by re-orientation of the O–H of O_3Cl–O–H and the H_2O molecules which thus result. The ClO–H proton is then passed along to another temporarily created H_2O. A quantum-mechanical calculation was later used to confirm the feasibility of this process in the crystal [93]. However, it may be difficult to obtain a sufficient weight of experimental evidence to confirm this.

The history of studies of the conductivity of ice is a long and tangled one, and a large amount of literature exists on this topic. This will not be covered in detail here except to mention the early papers by Gränicher [348] and Eigen and De Maeyer [349] and some very recent papers which should give the interested reader a lead into the current literature [350–352]. The area was also covered in the reviews by Glasser [344] and Franks [353]. The importance of H_3O^+ in the transport mechanism models has varied over the years, but a commonly accepted view seems to be that the mechanism depends on the formation of ionic defects H_3O^+ OH$^-$ (present in minute quantities), hopping of the H_3O^+ defect ion and the passage of Bjerrum L defects [354] which reorient the H_2O molecules (an L-defect is a vacant O–O where an H would normally reside). The most recent paper, however, presents a determination of the proton diffusion rate in doped ice [352], and the results led the authors to conclude that doped ice is in fact an insulator rather than a semiconductor. They hinted that the same is probably true for pure ice, observing that the early experiments drastically overestimated the proton transport. On the other hand, a recent low-temperature FTIR study of cubic ice [350] in which the conversion of D_2O to HOD molecules was monitored, found that ionic defects and L defects were roughly equally important charge carriers. Activation energy (E_a) values for hopping and L-defect migration of 39.7 kJ mol^{-1} and 50.2 kJ mol^{-1} respectively were obtained. A similar FTIR study of ethylene oxide (oxirane) clathrate hydrate [355] found a much smaller E_a of 20.9 kJ mol^{-1} for proton exchange than in cubic ice and that the majority charge carrier was the orientational L-defect, which was apparently fairly abundant. The small E_a was tentatively explained in terms of the difference in energy between formation of the L-defects $(= 32.2 \text{ kJ mol}^{-1})$ and the energy required to release protons trapped by the abundant L-defects $(= 53.1 \text{ kJ mol}^{-1})$.

3.7. Other Studies

$H_3O^+ClO_4^-$ is the most intensively studied oxonium compound (see appendix) and mention is made here of the calorimetric studies [89–91]. The H_3O^+ compound has a phase transition at 248.4 K [90], but it would appear that the D_3O^+ compound has two phase transitions; at 245.1 and 251.9 K [91]. A neutron powder diffraction study carried out at temperatures corresponding to these three phases suggested that the low-temperature phase and the intermediate phase have the same space group and very similar cell parameters, though there were intensity changes and slight shifts of the lines in the diffraction pattern [64]. Further investigation of this intermediate phase is necessary; it does not appear to have been detected in the ^2H NMR results [81]. In relation to this a point charge model has been used to attempt to simulate the rotational potential of the H_3O^+ ion [92] in the three phases.

Most of the theoretical ab-initio calculations involving the hydrated proton consider species isolated in space, and these will be discussed in Part Two, to be published in Water Science Reviews, Vol. 3. Two papers, however, have dealt with attempts to calculate the effects of external crystalline forces on the structure of the oxonium ion in solids [356, 257]. The results generally confirm the notion that the ion's conformation is distorted from its C_{3v} pyramidal shape most strongly by hydrogen bonding, with a tendency to lengthen the O–H bonds and decrease the H–O–H angles in order to form more nearly linear O–H···O bonds.

4. The Gaseous State

4.1. *Mass Spectrometry*

Ion-solvent molecule interactions and ion–molecule equilibria in the gas phase have been extensively studied by high-pressure mass spectrometry. From studies of the temperature dependence on the equilibrium constant K values of the thermodynamic variables, $\Delta H°$, $\Delta G°$ and $\Delta S°$ have been obtained. Information concerning the clustering of molecules around ions leads to a better understanding of the strong inner-shell solvation. The approach is described in a number of reviews [364–368].

The kinetics of the gas-phase reactions

$$H^+(H_2O)_{n-1} + H_2O + M = H^+(H_2O)_n + M$$

have been extensively studied [369–377]. In early measurements [370, 375] methane gas (M) at pressures of 3–4 Torr containing known pressures of water (1 to 200 mTorr) was passed through the ion source; ionization was obtained with a pulsed electron beam of 2000 V. Both CH_5^+ and $C_2H_5^+$ are formed and react with water to produce H_3O^+. These were subjected to mass analysis. Ethane, propane and neat water have also been used as major gases. The H_3O^+ forms higher hydrates by the reactions $n-1, n$:

$$H_3O^+ + H_2O + M = H^+(H_2O)_2 + M \qquad (1, 2)$$
$$H^+(H_2O)_2 + H_2O + M = H^+(H_2O)_3 + M \qquad (2, 3)$$
$$H^+(H_2O)_{n-1} + H_2O + M = H^+(H_2O)_n + M \qquad (n-1, n).$$

From measurements of the rate constants, at different temperatures (296–900 K) equilibrium constants, enthalpy and free energy values have been obtained.

Early investigations based on ion sampling from a field-free source at approximately atmospheric pressure [370, 371] led to heats of hydration which were in reasonable agreement with each other, with heats of reaction measured by a collisional detachment technique [378] and with theoretical calculations [379]. These results, particularly for the 1, 2 and 2, 3 reactions, differed markedly from those of Beggs and Field [372, 373] and Bennett and Field [374] and it was suggested that the disagreement arose because equilibrium had not been established [375], a view that was initially challenged [374] and later accepted [380]. Meisels *et al.* [381] demonstrated that equilibrium in proton hydration is not achieved within the average residence times of ions in chemical ionization sources. The analysis demonstrated that independence of pressure and linearity of van't Hoff plots do not assure the achievement of equilibrium and supported the thermodynamic parameters reported by Kebarle and co-workers [370, 375] and DePaz *et al.* [378]. Meisels *et al.* [382] have further discussed the assumptions and approximations underlying the use of time-resolved measurements.

Table 8. *Thermochemical data for the equilibrium*
$$H^+(H_2O)_{n-1} + H_2O = H^+(H_2O)_n$$

$n-1, n$	$-\Delta H^\circ_{n-1, n}$ (kJ mol^{-1})	$-\Delta G^\circ_{n-1, n}$ (kJ mol^{-1})	$-\Delta S^\circ_{n-1, n}$ (JK^{-1} mol^{-1})
0, 1	697†		
1, 2	132	102	102
2, 3	81.6	54.4	90.8
3, 4	74.9	39.7	119
4, 5	53.1	23.4	97.9
5, 6	48.5	17.2	105
6, 7	44.8	12.6	109.2
7, 8	41.8		

† Proton affinity of water [384].

Recently Kebarle and co-workers [376] have remeasured the $K_{n-1, n}$ values over the temperature range 80 to $-40\,°C$, because of concern that the use of neat water vapour for higher clusters with $n \geqslant 4$ had lead to high values of $-\Delta H^\circ_{n-1, n}$. Their results are summarized in table 8 and figure 14. The new $K_{3, 4}$ value was in excellent agreement with the early value of Cunningham *et al.* [375] (also compare Meot-Ner and Field [380]). Departure from the data of Kebarle *et al.* [370] at higher temperatures was noted and possible reasons discussed. The new results show a distinct break in the plot of $\Delta H^\circ_{n-1, n}$ when passing from $H_3O^+(H_2O)_3$ to $H_3O^+(H_2O)_4$ (i.e. $n-1 = 4$ to $n = 5$); the enthalpy change for adding the fourth water is 29 per cent less, compared with an 8.2 per cent drop between the second and third waters. This break was not evident in the earlier measurements [370]. The early value of $-\Delta H^\circ_{4, 5}$ was 62.8 kJ mol^{-1} compared with the new value of 53.1 kJ mol^{-1}. A similar break was observed for the NH_4^+ ion after four NH_3 groups had been added, thus completing the inner shell [383]. The accuracy of the value for $-\Delta H^\circ_{4, 5}$ is very important to structural inferences. The break is not as apparent in the $-\Delta G^\circ_{n-1, n}$ against n plot because of compensation by the $\Delta S^\circ_{n-1, n}$ values.

The results suggest that after the addition of one water molecule to H_3O^+, there is a marked fall-off in energy (38 per cent) suggesting the relative stability of $H_3O^+ \cdot H_2O$ (or $H_5O_2^+$). It has been suggested that the proton is equally strongly bonded to each of the surrounding water molecules in $H^+(H_2O)_2$ [370]. The new results further suggest that after the addition of three water molecules to H_3O^+ (formally giving $H_9O_4^+$) the fourth and subsequent water molecules are more weakly held (*cf.* [379]). This picture is consistent with the Eigen structure (figure 1).

The 0, 1 reaction corresponds to the process for the proton affinity of water. A comprehensive collection of gas-phase basicities and proton affinities,

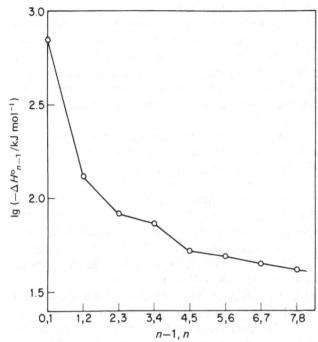

Figure 14. The change in $\Delta H^\circ_{n-1,\,n}$ (plotted as the logarithm) with increasing $n-1,\, n$ [372].

evaluated for internal consistency, has been presented by Lias *et al.* [384]. Data to June 1983 were included. They emphasize that for data derived from equilibrium constant measurements, absolute values of proton affinities depend on the proton-affinity values selected for a comparison standard and there has been considerable variation in this reference value. For water their selected value for the proton affinity is 697 ± 8 kJ mol^{-1}; the value of $\Delta H^\circ_f(H_3O^+)$ is 592 kJ mol^{-1}, according to the stationary electron convention. This value is in agreement with the appearance energy of H_3O^+ from a hydrogen-bonded dimer [385] and with an *ab-initio* calculation [386]. It is also consistent with the most recent studies of Collyer and McMahon [387] and McMahon and Kebarle [388].

The study of molecular clusters forms a link between isolated gas-phase species and the condensed phase. Studies of clusters also contribute to the understanding of the forces responsible for nucleation phenomena. Large stable water clusters have been detected in molecular-beam mass-spectrometer experiments and evidence has accumulated for a 'magic number' stability of the clusters, $(H_2O)_{21}\,H^+$ and $(D_2O)_{21}H^+$. In an early report by Lin [389] it was suggested that the result might be attributed to an experimental artefact but Searcy and Fenn [390] confirmed that a cluster distribution of remarkable stability existed for the cluster with $n=21$. They suggested a structure consisting of a pentagonal dodecahedron with a single oxygen atom at each corner, and with a single water molecule trapped inside. Kassner and Hagen [391] pointed out that this increased stability of the 21 molecule cluster had been predicted by molecular cluster theory. Distinct minima in the energy of formation occurred at those sizes corresponding to closed cage structures e.g. 20, 35, 47 – molecules for clusters whose cages are unoccupied and 21, 37, 50 – molecules for clusters all of whose cages are occupied with either a neutral molecule or an H_3O^+ ion.

Other groups have confirmed intensity anomalies corresponding to $n = 3$, 4, 21 and 28 of the species $(H_2O)_n H^+$ in mass spectra with a variety of techniques [392–399]. Lancaster et al. [392] subjected an ice surface to 0.5–3 keV He^+ ions and mass analysed the positive and negative secondary clusters in a quadrupole mass filter (SIMS). They observed that ion clusters up to $(H_2O)_{51} H^+$; $(H_2O)_4 H^+$ and $(H_2O)_{21} H^+$ had unusually high intensities thus indicating their particular stability. Several groups studied the electron impact ionization of neutral water clusters [395, 397–399]. Hermann et al. [397] suggested that a clathrate-like structure might be stabilized by a mobile 'surface' proton. Shinohara et al. [400, 401] found similar intensity irregularities in the mass spectral patterns of water–ammonia binary clusters at $(H_2O)_{20}(NH_3)_m H^+$ ($m = 1$–6) and showed that deformed pentagonal dodecahedral cages of 20 water molecules with an NH_4^+ ion trapped inside can account for the special stability of the magic number ions. This was supported by a Monte Carlo calculation. The stability of the magic number ions was found to be due to a strong Coulombic interaction between NH_4^+ ion and the surrounding 20 water molecules which form the dodecahedral cage structure.

This group extended this study [402]. Experimental evidence that both the electron-impact mass spectrum and the threshold vacuum-UV photoionization at 11.83 eV exhibit a distinct intensity drop between $(H_2O)_{21} H^+$ and $(H_2O)_{22} H^+$ and less distinct anomalies at $(H_2O)_{28} H^+$ and $(H_2O)_{30} H^+$ was presented along with results from further Monte Carlo simulations. The Monte Carlo calculations were carried out at the temperatures of 200, 150, 100, and 50 K for the ionized water clusters $(H_2O)_n H^+$ around $n = 21$ and also around $n = 28$. These clusters were found to have greater binding energies per molecule than their neighbours; the enhancement for $(H_2O)_{28} H^+$ was temperature-dependent and neutral interaction was larger than the ionic interaction whereas at $n = 20$ the ionic hydrogen bonding between H_3O^+ and H_2O was much larger than the neutral H_2O–H_2O interaction. Thus it was concluded that the 'strong ionic (electrostatic) interaction exerted between the encaged H_3O^+ ion and the surrounding water molecules plays a crucial role in stabilizing the ion–clathrate structure'. The enhanced stability of the $(H_2O)_{28} H^+$ ion was also attributed to an ion-centred cage structure composed of a cavity formed by 26 water molecules with an $(H_2O \cdot H_3O)^+$ dimer trapped inside ($H_5O_2^+$ (?)). The extent to which these conclusions bear on the secondary hydration of H_3O^+ in dilute aqueous solution is a matter for future study.

4.2. *Spectroscopic Studies*

Spectroscopic studies of the molecular ion H_3O^+ in the gas phase have, in part, been motivated by astronomical studies; it is calculated to be a very abundant molecular ion in the gas phase of interstellar clouds [403–406]. Both H_3O^+ and $H_5O_2^+$ have been detected at altitudes of 64–112 km [407]. The detection of H_3O^+ in space would be significant because it would be an indirect method of determining the abundance of H_2O, which itself has no low-lying transitions in the microwave or millimetre regions of the spectrum which can be excited under the conditions in interstellar space. It is also a key ion in the interstellar chemistry of OH and H_2O.

The first measurements of the infrared spectra of modes of H_3O^+ and $H_3O^+(H_2O)_n$ clusters in the gas phase were made by Schwarz [408]. Ions (about 5×10^{10} cm^{-3}) with life-times of several microseconds were generated by pulse radiolysis of argon containing small amounts of water vapour. Several strong and some weak broad bands were extracted from the spectra and assigned to H_3O^+, $H_9O_4^+$, $H_{11}O_5^+$ and $H_{13}O_6^+$. *Ab-initio* calculations of the hydrated H_3O^+ ion, giving molecular structures and force constants for OH stretching modes, (and a wealth of related information for the higher hydrates) were made by Newton [409]. These were used to substantiate the assignments of the observed infrared band frequencies. The calculated and experimental values are collected in table 9. The model is similar to that illustrated by figure 1. After the addition of three water molecules to the H_3O^+, the fourth is added to the second coordination sphere. The hydronium oxygen does not act as a proton acceptor; appreciable repulsion is inferred from a calculation with this model. Thus for $H_{11}O_5^+$ one water molecule in the first coordination sphere of H_3O^+ has one O–H bond which is not hydrogen bonded and to this was assigned a frequency similar to that of $H_9O_4^+$ (i.e. 3600–3700 cm^{-1}); it has a second O–H bond which is weakly hydrogen bonded to the water molecule in the second coordination sphere and to this was assigned a relatively high frequency band (3200 cm^{-1}). The OH bond of the central H_3O^+, which is strongly hydrogen bonded to this water molecule, will be weakened and thus a diffuse low-frequency band was assigned to it (2180 cm^{-1}). Of the other two OH stretching vibrations of the central H_3O^+ unit, one was assigned to the 2860 cm^{-1} band and the other was ascribed to either an asymmetry of this band or to low intensity which was not discriminated from the noise. It is interesting to note that the order $\nu_{3,\,asym} > \nu_{1,\,sym}$ for H_3O^+ switches to $\nu_{3,\,asym} < \nu_{1,\,sym}$ for $H_3O^+ (H_2O)_3$.

The high-resolution infrared spectrum of the ν_3, doubly-degenerate, antisymmetric stretch of H_3O^+ has been recorded by Begemann *et al.* [410]. Excitation of this vibration involves a non-zero transition dipole moment

Table 9. Infrared frequencies and intensities of the OH stretching modes

Species	Mode	Frequency (cm^{-1}) Calc.	Exp.	Relative intensity
H$_2$O	ν_3 asym	3750	3756	0.38
	ν_1 sym	3653	3652	0.07
(H$_3$O)$^+$(H$_2$O)$_n$				
n = 0	ν_3 asym	3496	3490	13.0
	ν_1 sym	3403	—	2.0
n = 3	ν_3 asym	2682	3000 sh	55
	ν_1 sym	2846	2660 s	6.3
n = 4	asym	2844	2860 s	48
	sym	2918	(2900–3000)^a	21
	b	2173	2180 w, br	55
	c	3307	3200 m	—
n = 5	—	3106	—	—
		2477	—	—
		—	3340 m	—

s strong; m, medium; w, weak; sh, shoulder; br, broad.
a Region of apparent asymmetry in the 2860 cm^{-1} peak.
b The weakest O–H bond of H$_3$O$^+$, being bonded to that water which in turn is bonded to the fourth (outer) water molecule.
c That O–H stretch of the intermediate water molecule which is involved in hydrogen bonding to the outer water molecule.

perpendicular to the symmetry axis of the molecule and thus the spectrum exhibits the features of a perpendicular-type band and features attributable to the inversion doubling of all states. An analysis of 60 transitions in the symmetric and anti-symmetric inversion states were observed and analysed; the effective band-origins were found to be 3530.165 (55) and 3513.840 (47) cm^{-1} respectively (cf. Schwarz, 3490 cm^{-1}). The O–H bond length was calculated as 97.9 (6) pm and the H–O–H angle was 114.91 (45)°. Other molecular constants were reported.

Bunker et al. [411] also studied the $\nu_3(0^+ \leftarrow 0^+)$ and $\nu_3(0^- \leftarrow 0^-)$ bands of H$_3$O$^+$ using a difference frequency spectrometer. Forty-one observed lines were fitted (including lines of the $\nu_2 (1^- \leftarrow 0^+)$ inversion mode; Haese and Oka [412]) and rotational constants and inversion energies were calculated. The following parameters were given: bond length r_e 97.6 pm; the H–O–H angle 110.7°; the height of the barrier to inversion $V(r_p, 120°) - V(r_e, \alpha_e)$, 904 cm^{-1}; and the lowering of the potential at the planar symmetric configuration when the bond length is reduced from r_e to r_p, 32 cm^{-1}.

Haese and Oka [412] and Liu et al. [414] observed the high resolution ν_2 $(1^- \leftarrow 0^+)$ and $(1^+ \leftarrow 0^-)$ inversion bands of H$_3$O$^+$ in absorption using diode laser spectroscopy. The two-band origins were found to be at 954.4003 (25) and 525.8237 (13) cm^{-1} respectively, in agreement with the prediction of

Botschwina *et al.* [413]. The ground-state rotational B value was 11.2537 (6) cm^{-1} in good agreement with the value 11.23 cm^{-1} from the ν_3 band observed by Begemann *et al.* [410]. Liu and Oka [415] have recently observed the $1^- \leftarrow 1^+$ separation, 373.2304 (47) cm^{-1}, and have thus fixed the ground-state splitting $(0^- \leftarrow 0^+)$ as 55.3462 (55) cm^{-1}. Thus the parameters of the double-minimum potential well are being mapped in detail.

Davies *et al.* [416] have also detected more than 100 absorption lines of H_3O^+ in an A.C. discharge by using the velocity-modulation technique. Many lines were assigned to the $1^- \leftarrow 0^+$ and $1^+ \leftarrow 0^-$ components of the ν_2 inversion mode; band centres were 954.40201 (186) and 525.8298 (63) cm^{-1} respectively. Molecular constants are given.

Lemoine and Destombes [417] have used a longitudinal magnetic field to enhance the concentration of ions in a glow discharge and have observed new transitions corresponding to high rotational levels of the H_3O^+ $(1^- \leftarrow 0^+)$ band. The intensity enhancement is about two orders of magnitude. ν_0 was calculated to be 954.428 (14) cm^{-1}.

Sears *et al.* [418] have recently reported the observation and assignment of vibration–rotation transitions in the fully deuterated D_3O^+. These data lead to the determination of the zero-point and equilibrium structure of the ion and to a reliable determination of the inversion potential function. The experiments have benefited by the guidance from the precision of modern *ab-initio* studies. The theoretical estimates of vibrational band origins have directed the experimentalist to detect more quickly the spectral transitions by tuning the infrared laser sources to the correct spectral region. Complete analysis of the $(1^- \leftarrow 0^+)$ band was achieved and, with more difficulty, many lines of the lower frequency $(1^+ \leftarrow 0^-)$ band. The band origins were 645.13042 (41) and 438.3930 (12) cm^{-1} respectively. Fits to both the symmetric top model and the non-rigid invertor model [419] were performed. The equilibrium geometry was derived: $r_e = 97.58$ pm; $\alpha_e(\text{H–O–H}) = 111.3°$; $r_0 = 98.33$ pm; $\alpha_0(\text{H–O–H}) = 116.4°$ (zero-point structure); the non-rigid invertor model was fitted to all extant H_3O^+ and D_3O^+ spectroscopic data. The barrier height is 672 cm^{-1}. The inversion potential diagram is shown in figure 15.

Plummer *et al.* [420] have detected an absorption frequency at 307192.41 (5) MHz which they believe to be the P(2, 1) transition of H_3O^+ $(J, K = 1, 1 \leftarrow 2, 1)$, predicted to occur at 307 167 (240) MHz [415]. Bogey *et al.* [421] have measured the $K = 0, 1, 2$ components of the $J = 2^- \rightarrow 3^+$ transitions. The basic information necessary for astrophysical searches with radioastronomical receivers is now available.

Further results for partially deuterated H_3O^+ and for clusters $H_3O^+(H_2O)_n$ can be anticipated from diode laser techniques. Thus, in the absence of bulk solvent, the molecular parameters of H_3O^+, D_3O^+ and their higher hydrates will certainly become known. At this time the $H_5O_2^+$ species in the gas phase has not been characterized by spectroscopy. It is interesting to note that the

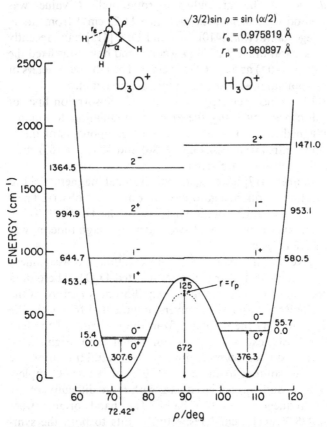

Figure 15. The inversion potential for the oxonium ion (with $r = r_e$ fixed) showing the positions of the lowest few inversion states for H_3O^+ and D_3O^+. (Reproduced, with permission, from the American Institute of Physics and P. R. Bunker from reference 418.)

bond length and bond angle of H_3O^+ in the gas phase is similar to that found for crystalline $H_3O^+CF_3SO_3^-$ (figure 2) (compare 98.33 pm and 116.4° with 98.6, 99.7 and 99.8 pm and 116.0, 111.7 and 110.9°) but it is significantly shorter than that found for H_3O^+ para-$CH_3(C_6H_4)SO_3^-$ (figure 3).

5. Summary

We have reviewed the literature concerned with the nature of the hydrated proton as found in the two extreme states of matter, the solid and gaseous states. In both states the species have been characterized very well; the existence and shape of a well defined H_3O^+ entity have been clearly established. Higher hydrates also exist; H_3O^+ and $H_5O_2^+$ appear to be the basic units and thus all hydrates may be described in terms of $H_3O^+ \cdot nH_2O$ or $H_5O_2^+ \cdot nH_2O$. In some cases $H_5O_2^+$ might better be described as

H_3O^+. H_2O, if one accepts the criteria set out earlier. Hydrogen bonding plays an extremely important role in determining both the formation of the higher hydrates and the conformation of all the species in almost all solids. The lattice environment also plays an important role in determining whether a particular crystal is an ionogen or an ionophore.

The knowledge gained from these solid and gas phase studies is invaluable for attempts to understand the much more intricate liquid phase and, in particular, aqueous solutions. The literature for this phase, where the interactions with neighbouring molecules and ions vary rapidly as a function of time and hence the lability of the proton must be considered, as well as the stability of a particular species, will be reviewed in Part Two. The current state of theoretical studies will also be addressed.

Acknowledgments

The authors express their thanks to Dr N. Taylor and Dr C. Chieh for assistance in the preparation of figures 2–5, to Dr A. Schultz, Dr J. M. Williams and Dr P. R. Bunker for helpful correspondence, to Dr T. B. McMahon and Dr W. A. E. McBryde for helpful discussion and to Mrs Lisa Dowsett for assistance with the manuscript. This work was supported by a grant from the Natural Sciences and Engineering Research Council of Canada. It is published as NRCC 25948.

References

1. A. S. Eve, *Rutherford*, Cambridge University Press, 1939, p. 43.
2. P. Walden, *Salts, Acids and Bases: Electrolytes: Stereochemistry*, McGraw-Hill, New York, 1929, p. 60 ff.
3. R. P. Bell, *The Proton in Chemistry*, Cornell University Press, Ithaca, New York, 1959; Second edn: Chapman and Hall, London, 1973.
4. W. B. Jensen, *The Lewis Acid-Base Concepts, An Overview*, Wiley-Interscience: New York, 1980.
5. R. P. Bell, *Chem. Soc. Quart. Rev.* **1** 113 (1947).
6. E. N. da C. Andrade, *Rutherford and the Nature of the Atom*, Doubleday, New York, 1964, p. 164. The termination -on dates back to M. Faraday and W. Whewell and the words ion, cation and anion. S. Ross. *Notes and Records of R. Soc.*, *London* **16**, 187 (1961).
7. J. D. Stranathan, *The Particles of Modern Physics*, Blakiston, New York, 1942, chap. 5.
8. *Pure Appl. Chem.* **51**, 35 (1979).
9. *Tables of Interatomic Distances and Configurations in Molecules and Ions*, Spec. Publ. No. 18 (L. E. Sutton, ed.), The Chemical Society, London, 1965.
10. J. N. Brönsted, *Rec. Trav. Chim. Pays-Bas* **42**, 718 (1923).
11. The contribution of T. M. Lowry (*Chem. Ind.* (*London*) **42**, 43 (1923)) does not provide as explicit a definition of the acid. See reference 3, p. 7, footnote 2.
12. R. J. Gillespie, in *Proton Transfer Reactions* (E. Caldin and V. Gold, eds), Chapman and Hall, London, 1975, p. 1.

13. G. N. Lewis, *Valency and the Structure of Atoms and Molecules*, The Chemical Catalog Co, New York, 1923.
14. J. Bjerrum, *Angew. Cheml.* **63**, 527 (1951); *Naturwissenschaft.* **38**, 461 (1951).
15. M. Volmer, *Justus Liebigs Ann. Chem.* **440**, 200 (1924).
16. R. E. Richards & J. A. S. Smith, *Trans. Faraday Soc.* **47**, 1261 (1951).
17. Y. Kakiuchi, H. Shono, H. Komatsu & K. Kigoshi, *J. Chem. Phys.* **19**, 1069 (1951).
18. H. Goldschmidt & O. Udby, *Z. Phys. Chem.* **60**, 728 (1907).
19. A. Hantzsch, *Z. Phys. Chem.* **61**, 257 (1908); *Z. Elektrochem.* **29**, 230 (1923).
20. K. Fajans & G. Joos, *Z. Phys.* **23**, 1 (1924).
21. K. Fajans, *Ber. d. Deutschen Phys. Gesell.* **21**, 709 (1919).
22. E. Wicke, M. Eigen & Th. Ackermann, *Z. Phys. Chem., N.F.* **1**, 340 (1954).
23. M. Eigen & L. De Maeyer, in *The Structure of Electrolytic Solutions* (W. J. Hamer, ed.), Wiley, New York, 1959, chap. 5, p. 64.
24. *Nomenclature of Inorganic Chemistry, 2nd edn, Definitive Rules 1970 (Pure and Appl. Chem.* **28**, 1971), Butterworths, London p. 20–1.
25. *J. Am. Chem. Soc.* **82**, 5530 (1960).
26. R. M. Fuoss, *J. Chem. Educ.* **32**, 527 (1955).
27. P. A. Giguère, *Rev. Chim. Mineral.* **3**, 627 (1966).
28. J. M. Williams, *National Bureau of Standards (US), Spec. Publ.* **301**, 237 (1969).
29. J.-O. Lundgren, *Acta Univ. Upsal.* **271**, 1 (1974).
30. J. Emsley, *Chem. Soc. Rev.* **9**, 91 (1980).
31. I. Taesler, *Acta Univ. Upsal.* **591**, 1 (1981).
32. J. Rozière, 'Spectroscopie de Vibrations des Hydrates du Proton dans les Cristaux', PhD Thesis, Montpellier, 1973.
33. J.-O. Lundgren & I. Olovsson, in *The Hydrogen Bond, Vol. II* (P. Schuster, G. Zundel & C. Sandorfy, eds), North-Holland, New York, 1976, chap. 10, p. 471.
34. J. M. Williams, in *The Hydrogen Bond, Vol. II* (P. Schuster, G. Zundel & C. Sandorfy, eds), North-Holland, New York, 1976, chap. 14, p. 655.
35. J. M. Williams & S. W. Peterson, in *Spectroscopy in Inorganic Chemistry, Vol. II*, Academic Press, New York, 1971.
36. P. A. Giguère, *J. Chem. Educ.* **56**, 571 (1979).
37. H. L. Clever, *J. Chem. Educ.* **40**, 637 (1963).
38. D. Mootz, U. Ohms & W. Poll, *Z. Anorg. Allg. Chem.* **479**, 75 (1981).
39. C. C. Ferriso & D. F. Hornig, *J. Chem. Phys.* **23**, 1464 (1955).
40. J. T. Mullhaupt, 'The Vibrational Spectrum of the Hydronium Ion in Crystals', PhD Thesis, University Rochester, 1958.
41. D. Mootz & W. Poll, *Z. Anorg. Allg. Chem.* **484**, 158 (1982).
42. Y. K. Yoon & G. B. Carpenter, *Acta Crystallogr.* **12**, 17 (1959).
43. C. C. Ferriso & D. F. Hornig, *J. Am. Chem. Soc.* **75**, 4113 (1953).
44. A. S. Gilbert & N. Sheppard, *J. Chem. Soc. Chem. Commun.* **337** (1971).
45. A. S. Gilbert & N. Sheppard, *J. Chem. Soc. Faraday Trans. 2* **69**, 1628 (1973).
46. B. Desbat & P. V Huong, *Spectrochim. Acta* **31A**, 1109 (1975).
47. T. Huston, I. C. Hisatsune & J. Heicklen, *Can. J. Chem.* **61**, 2077 (1983).
48. J.-O. Lundgren & I. Olovsson, *Acta Crystallogr.* **23**, 966 (1967).
49. J. Rozière & J. Potier, *J. Mol. Struct.* **13**, 91 (1972).
50. J.-O. Lundgren & I. Olovsson, *Acta Crystallogr.* **23**, 971 (1967).

51. I. Taesler & J.-O. Lundgren, *Acta Crystallogr.* **B34**, 2424 (1978).
52. J.-O. Lundgren, *Acta Crystallogr.* **B26**, 1893 (1970).
53. R. Attig & J. M. Williams, *Angew. Chem. Int. Ed. Engl.* **15**, 491 (1976).
54. M. H. Herzog-Cance, J. Potier & A. Potier, *Adv. Mol. Relax. Int. Processes* **18**, 31 (1980).
55. J.-O. Lundgren & I. Olovsson, *J. Chem. Phys.* **49**, 1068 (1968).
56. J. Rudolph & H. Zimmermann, *Z. Phys. Chem. N.F.* **43**, 311 (1964).
57. A. G. Maki & R. West, *Inorg. Chem.* **2**, 657 (1963).
58. L. W. Schroeder & J. A. Ibers, *J. Am. Chem. Soc.* **88**, 2601 (1966).
59. L. W. Schroeder & J. A. Ibers, *Inorg. Chem.* **7**, 594 (1968).
60. F. S. Lee & G. B. Carpenter, *J. Phys. Chem.* **63**, 279 (1959).
61. M. R. Truter, *Acta Crystallogr.* **14**, 318 (1961).
62. C. E. Nordman, *Acta Crystallogr.* **15**, 18 (1962).
63. H. G. Smith & H. A. Levy, *Abstracts from American Crystallography Association, Abs. 41, (1959)*, American Crystallography Association, Ithaca, New York.
64. J. Domoslawski & M. Golab, *Physica B* **101**, 217 (1980).
65. D. E. Bethell & N. Sheppard, *J. Chem. Phys.* **21**, 1421 (1953).
66. D. J. Millen & E. G. Vaal, *J. Chem. Soc.* 2913 (1956).
67. R. C. Taylor & G. L. Vidale, *J. Am. Chem. Soc.* **78**, 5999 (1956).
68. J. T. Mullhaupt & D. F. Hornig, *J. Chem. Phys.* **24**, 169 (1956).
69. R. Savoie & P. A. Giguère, *J. Chem. Phys.* **41**, 2698 (1964).
70. J. A. Janik, J. M. Janik, J. Mellor & H. Palevsky, *J. Phys. Chem. Solids* **25**, 1091 (1964).
71. J. Janik, *Acta Phys. Pol.* **27**, 491 (1965).
72. J. M. Janik, J. A. Janik, A. Bajorek & K. Parliński, *Phys. Status. Solidi* **9**, 905 (1965).
73. M. Fournier, G. Mascherpa, D. Rousselet & J. Potier, *C. R. Acad. Sci. Paris C* **269**, 279 (1969).
74. M. Fournier & J. Rozière, *C.R. Acad. Sci. Paris C* **270**, 729 (1970).
75. J. M. Janik, G. Pytasz, M. Rachwalska, J. A. Janik, I. Natkaniec & W. Nawrocik, *Acta Phys. Pol. A* **43**, 419 (1973).
76. J. A. Janik, in *The Hydrogen Bond, Vol. III* (P. Schuster, G. Zundel & C. Sandorfy, eds), North-Holland, New York, 1976, chap. 19, pp. 903, 915.
77. C. I. Ratcliffe & D. E. Irish, *Can. J. Chem.* **62**, 1134 (1984).
78. M. P. Thi, M.-H. Herzog-Cance, A. Potier & J. Potier, *J. Raman Spectrosc.* **12**, 238 (1982).
79. Y. Kakiuchi, H. Shono, H. Komatsu & K. Kigoshi, *J. Phys. Soc. Japan* **7**, 102 (1952).
80. J. W. Hennel & M. Pollack-Stachura, *Acta Phys. Pol.* **35**, 239 (1969).
81. D. E. O'Reilly, E. M. Peterson & J. M. Williams, *J. Chem. Phys.* **54**, 96 (1971).
82. M.-H. Cance & A. Potier, *J. Chim. Phys.–Phys. Chim. Biol.* **68**, 941 (1971).
83. M. H. Herzog-Cance, J. Potier & A. Potier, *Adv. Mol. Relax. Int. Processes* **14**, 245 (1979).
84. C. I. Ratcliffe, J. A. Ripmeester & J. S. Tse, *Chem. Phys. Lett.* **120**, 427 (1985).
85. Y. Kakiuchi & H. Komatsu, *J. Phys. Soc. Japan* **7**, 380 (1952).
86. E. R. Andrew & N. D. Finch, *Proc. Phys. Soc. (London), B* **70**, 980 (1957).
87. A. B. Yaroslavtsev, V. F. Chuvaev, Z. N. Prozorovskaya & I. I. Baskin, *Zh. Neorg. Khim.* **28**, 2746 (1983).

88. C. I. Ratcliffe, unpublished results (1985).
89. J. B. Rosolovski & A. A. Zinoviev, *Zh. Neorg. Khim.* **3**, 1589 (1958).
90. J. M. Janik, M. Rachwalska & J. A. Janik, *Physica* **72**, 168 (1974).
91. K. Czarniecki, J. A. Janik, J. M. Janik, G. Pytasz, M. Rachwalska & T. Waluga, *Physica B* **85**, 291 (1977).
92. K. Czarniecki, *Physica B* **103**, 226 (1981).
93. J. Angyan, M. Allavena, M. Picard, A. Potier & O. Tapia. *J. Chem. Phys.* **77**, 4723 (1982).
94. A. Potier & D. Rousselet, *J. Chim. Phys. Phys.-Chim. Biol.* **70**, 873 (1973).
95. I. Olovsson, *J. Chem. Phys.* **49**, 1063 (1968).
96. A. C. Pavia & P. A. Giguère, *J. Chem. Phys.* **52**, 3551 (1970).
97. E. Chemouni, M. Fournier, J. Rozière & J. Potier, *J. Chim. Phys. Phys.-Chim. Biol* **67**, 517 (1970).
98. J. Almlöf, J.-O. Lundgren & I. Olovsson, *Acta Crystallogr. B* **27**, 898 (1971).
99. J. Almlöf, *Acta Crystallogr. B* **28**, 481 (1972).
100. J. Rozière & J. Potier, *J. Inorg. Nucl. Chem.* **35**, 1179 (1973).
101. M. H. Herzog-Cance, J. Potier & A. Potier, *Adv. Mol. Relax. Int. Processes* **20**, 165 (1981).
102. J. Almlöf, *Chem. Scr.* **3**, 73 (1973).
103. M. Pham Thi, M. H. Herzog-Cance, A. Potier & J. Potier, *J. Raman Spectrosc.* **11**, 96 (1981).
104. V. Luzzati, *Acta Crystallogr.* **4**, 239 (1951).
105. R. G. Delaplane, I. Taesler & I. Olovsson, *Acta Crystallogr. B* **31**, 1486 (1975).
106. D. E. Bethell & N. Sheppard, *J. Chim. Phys. Phys.-Chim. Biol.* **50**, C72 (1953).
107. J. M. Janik, J. A. Janik, A. Bajorek, K. Parliński & M. Sudnik-Hrynkiewicz, *Physica* **35**, 457 (1967).
108. M. H. Herzog-Cance, J. Potier, A. Potier, P. Dhamelincourt, B. Sombret & F. Wallart, *J. Raman Spectrosc.* **7**, 303 (1978).
109. V. Luzzati, *Acta Crystallogr.* **6**, 152 (1953).
110. V. Luzzati, *Acta Crystallogr.* **6**, 157 (1953).
111. I. Taesler, R. G. Delaplane & I. Olovsson, *Acta Crystallogr. B* **31**, 1489 (1975).
112. P. Bourre-Maladière, *C.R. Acad. Sci. Paris* **246**, 1063 (1958).
113. I. Taesler & I. Olovsson, *Acta Crystallogr. B* **24**, 299 (1968).
114. P. A. Giguère & R. Savoie, *Can. J. Chem.* **38**, 2467 (1960).
115. I. J. F. Poplett, *J. Magn. Reson.* **44**, 488 (1981).
116. M. Pollak-Stachura & S. Sagnowski, *Proc. XVIIth Cong. Ampère, Turku 1972*, (V. Hovi, ed.), North-Holland, London, 1973, p. 225.
117. I. Taesler & I. Olovsson, *J. Chem. Phys.* **51**, 4213 (1969).
118. J. A. S. Smith & R. E. Richards, *Trans. Faraday Soc.* **48**, 307 (1952).
119. T. Kjällman & I. Olovsson, *Acta Crystallogr. B* **28**, 1692 (1972).
120. D. Mootz, Institut für Anorganische Chemie und Struktur Chemie, Universität Düsseldorf, FRG, private communication, work to be published.
121. J.-O. Lundgren & I. Taesler, *Acta Crystallogr. B* **35**, 2384 (1979).
122. B. Krebs & M. Hein, *Z. Naturforsch. B* **34**, 1666 (1979).
123. R. Ripan & R. Palade, *Ann. Sci. Univ. Jassy, I*, **30**, 155 (1948).
124. R. G. Delaplane, J.-O. Lundgren & I. Olovsson, *Acta Crystallogr. B* **31**, 2208 (1975).
125. J. B. Spencer & J.-O. Lundgren, *Acta Crystallogr. B* **29**, 1923 (1973).

126. J.-O. Lundgren, R. Tellgren & I. Olovsson, *Acta Crystallogr. B* **34**, 2945 (1978).
127. R. G. Delaplane, J.-O. Lundgren & I. Olovsson, *Acta Crystallogr. B* **31**, 2202 (1975).
128. J.-O. Lundgren, *Acta Crystallogr. B* **34**, 2428 (1978).
129. J.-O. Lundgren, *Acta Crystallogr. B* **34**, 2432 (1978).
130. D. Mootz & H. Altenburg, *Acta Crystallogr. B* **27**, 1520 (1971).
131. H. Remy and H. Falius, *Naturwissenschaften* **43**, 177 (1956).
132. R. Attig & D. Mootz, *Z. Anorg. Allg. Chem.* **419**, 139 (1976).
133. D. E. C. Corbridge, *Acta Crystallogr.* **6**, 104 (1953).
134. T. Migchelsen, R. Olthof & A. Vos, *Acta Crystallogr.* **19**, 603 (1965).
135. R. Attig & D. Mootz, *Acta Crystallogr. B* **33**, 605 (1977).
136. H. Steinfink & G. D. Brunton, *Inorg. Chem.* **9**, 2112 (1970).
137. R. R. Ryan & R. A. Penneman, *Inorg. Chem.* **10**, 2637 (1971).
138. G. D. Brunton & C. K. Johnson, *J. Chem. Phys.* **62**, 3797 (1975).
139. J. B. Bates & L. M. Toth, *J. Chem. Phys.* **61**, 129 (1974).
140. G. M. Brown, M.-R. Noe-Spirlet, W. R. Busing & H. A. Levy, *Acta Crystallogr. B* **33**, 1038 (1977).
141. T. Gustafsson, J.-O. Lundgren & I. Olovsson, *Acta Crystallogr. B* **33**, 2373 (1977).
142. T. Gustafsson, J.-O. Lundgren & I. Olovsson, *Acta Crystallogr. B* **36**, 1323 (1980).
143. J. Tudo, B. Jolibois, G. Laplace & G. Nowogrocki, *Acta Crystallogr. B* **35**, 1580 (1979).
144. A. B. Yaroslavtsev, Z. N. Prozorovskaya, V. F. Chuvaev & V. I. Spitsyn, *Zh. Neorg. Khim.* **28**, 2495 (1983).
145. F. A. Cotton, C. K. Fair, G. E. Lewis, G. N. Mott, F. K. Ross, A. J. Schultz & J. M. Williams, *J. Am. Chem. Soc.* **106**, 5319 (1984).
146. G. Johansson, *Ark. Kemi.* **20**, 343 (1962).
147. D. W. Kydon, M. Pintar & H. E. Petch, *J. Chem. Phys.* **48**, 5348 (1968).
148. R. T. Thompson & D. W. Kydon, *J. Chem. Phys.* **61**, 1813 (1974).
149. R. C. L. Mooney-Slater, *Acta Crystallogr.* **14**, 1140 (1961).
150. M. G. Shilton & A. T. Howe, *J. Solid State Chem.* **34**, 137 (1980).
151. B. Morosin, *Acta Crystallogr. B* **34**, 3732 (1978).
152. C. Poinsignon, A. Fitch & B. E. F. Fender, *Solid State Ionics* **9, 10**, 1049 (1983).
153. G. J. Kearley, A. N. Fitch & B. E. F. Fender, *J. Mol. Struct.* **125**, 229 (1984).
154. Ph. Colomban, M. Pham Thi & A. Novak, *Solid State Commun.* **53**, 747 (1985).
155. M. Pham-Thi, Ph. Colomban & A. Novak, *J. Phys. Chem. Solids* **46**, 565 (1985).
156. P. E. Childs, T. K. Halstead, A. T. Howe & M. G. Shilton, *Mater. Res. Bull.* **13**, 609 (1978).
157. T. K. Halstead, N. Boden, L. D. Clark & C. G. Clarke, *J. Solid State Chem.* **47**, 225 (1983).
158. H. Kahil, M. Forestier, J. Guitton & J. P. Cohen-Addad, *Surf. Technol.* **26**, 1 (1985).
159. M. G. Shilton & A. T. Howe, *Mater. Res. Bull.* **12**, 701 (1977).
160. A. T. Howe & M. G. Shilton, *J. Solid State Chem.* **28**, 345 (1979).
161. A. T. Howe & M. G. Shilton, *J. Solid State Chem.* **34**, 149 (1980).
162. S. B. Lyon & D. J. Fray, *Solid State Ionics*, **15**, 21 (1985).
163. B. Morosin, *Phys. Lett. A* **65**, 53 (1978).

164. L. Bernard, A. N. Fitch, A. T. Howe, A. F. Wright & B. E. F. Fender, *J. Chem. Soc. Chem. Commun.* 784 (1981).

165. A. N. Fitch, A. F. Wright & B. E. F. Fender, *Acta Crystallogr. B* **38**, 2546 (1982).

166. A. N. Fitch, L. Bernard, A. T. Howe, A. F. Wright & B. E. F. Fender, *Acta Crystallogr. C* **39**, 159 (1983).

167. K. D. Kreuer, A. Rabenau & R. Messer, *Appl. Phys. A* **32**, 45 (1983).

168. K. D. Kreuer, A. Rabenau & R. Messer, *Appl. Phys. A* **32**, 155 (1983).

169. A. Nakahara, Y. Saito & H. Kuroya, *Bull. Chem. Soc. Japan* **25**, 331 (1952).

170. J. M. Williams, *Inorg. Nucl. Chem. Lett.* **3**, 297 (1967).

171. J. Rozière & J. M. Williams, *Inorg. Chem.* **15**, 1174 (1976).

172. R. D. Gillard & G. Wilkinson, *J. Chem. Soc.* 1640 (1964).

173. H. E. LeMay, Jr *Inorg. Chem.* **7**, 2531 (1968).

174. S. Ooi, Y. Komiyama, Y. Saito & H. Kuroya, *Bull. Chem. Soc. Japan* **32**, 263 (1959).

175. S. Ooi, Y. Komiyama & H. Kuroya, *Bull. Chem. Soc. Japan* **33**, 354 (1960).

176. Y. Saito & H. Iwasaki, 'Advances in the Chemistry of the Coordination Compounds', *Proc. 6th Int. Conf. Coord. Chem., Detroit* (S. Kirschner, ed.), Macmillan, New York, 1961, p. 557.

177. Y. Saito & H. Iwasaki, *Bull. Chem. Soc. Japan* **35**, 1131 (1962).

178. C. Dorémieux-Morin & E. Freund, *Bull. Soc. Chim. Fr.* **2**, 418 (1973).

179. D. Mootz & E. Oellers, *Z. Kristallogr.* **159**, 95 (1982).

180. D. Mootz & M. Steffen, *Z. Kristallogr.* **154**, 306 (1981).

181. Von D. Mootz & M. Steffen, *Z. Anorg. Allg. Chem.* **482**, 193 (1981).

182. H. Selig, W. A. Sunder, F. C. Schilling & W. E. Falconer, *J. Fluorine Chem.* **11**, 629 (1978).

183. H. Selig, W. A. Sunder, F. A. Disalvo & W. E. Falconer, *J. Fluorine Chem.* **11**, 39 (1978).

184. S. Cohen, H. Selig & R. Gut, *J. Fluorine Chem.* **20**, 349 (1982).

185. K. O. Christe, W. W. Wilson & C. J. Schack, *J. Fluorine Chem.* **11**, 71 (1978).

186. J. P. Masson, J. P. Desmoulin, P. Charpin & R. Bougon, *Inorg. Chem.* **15**, 2529 (1976).

187. K. O. Christe, P. Charpin, E. Soulie, R. Bougon, J. Fawcett & D. R. Russell, *Inorg. Chem.* **23**, 3756 (1984).

188. K. O. Christe, C. J. Schack & R. D. Wilson, *Inorg. Chem.* **14**, 2224 (1975).

189. B. Bonnet, J. Rozière, R. Fourcade & G. Mascherpa, *Can. J. Chem.* **52**, 2077 (1974).

190. U. Greher & A. Schmidt, *Z. Anorg. Allg. Chem.* **444**, 97 (1978).

191. H. S. Rade & D. Heitkamp, *Rev. Brasil. Fis.* **5**, 305 (1975).

192. J. Coetzer, W. Robb & P. v Z. Bekker, *Acta Crystallogr. B* **28**, 3587 (1972).

193. J. E. Fergusson & R. R. Sherlock, *Aust. J. Chem.* **30**, 1445 (1977).

194. H. Okumara, T. Taga, K. Osaki & I. Tsujikawa, *Bull. Chem. Soc. Japan* **55**, 307 (1982).

195. A. Bino & F. A. Cotton, *J. Am. Chem. Soc.* **101**, 4150 (1979).

196. H. Follner, *Acta Crystallogr. B* **26**, 1544 (1970).

197. C.-O. Selenius & R. G. Delaplane, *Acta Crystallogr. B* **34**, 1330 (1978).

198. R. J. Bateman & L. R. Bateman, *J. Am. Chem. Soc.* **94**, 1130 (1972).

199. J. M. Williams & S. W. Peterson, *J. Am. Chem. Soc.* **91**, 776 (1969).

200. D. E. O'Reilly, E. M. Peterson, C. E. Scheie & J. M. Williams, *J. Chem. Phys.* **55**, 5629 (1971).
201. H. J. Keller, I. Leichtert, G. Uhlmann & J. Weiss, *Chem. Ber.* **110**, 1684 (1977).
202. R. A. Penneman & R. R. Ryan, *Acta Crystallogr. B* **28**, 1629 (1972).
203. J. M. Smith, L. H. Jones, I. K. Kressin & R. A. Penneman, *Inorg. Chem.* **4**, 369 (1965).
204. G. Picotin, J. Rozière, J. Potier & A. Potier, *Adv. Mol. Relax. Int. Processes* **7**, 177 (1975).
205. P. V. Huong & B. Desbat, *J. Raman Spectrosc.* **2**, 373 (1974).
206. H. M. Matheson & W. A. Whitla, *Can. J. Chem.* **56**, 957 (1978).
207. R. Ortwein & A. Schmidt, *Z. Anorg. Allg. Chem.* **425**, 10 (1976).
208. G. Picotin & J. Rozière, *J. Chim. Phys. Phys.–Chim. Biol.* **69**, 372 (1972).
209. H. Henke, *Acta Crystallogr. B* **36**, 2001 (1980).
210. H. Henke, *Abstracts from the 5th European Crystallography Meeting, Copenhagen (1979)*.
211. R. Ortwein & A. Schmidt, *Z. Anorg. Allg. Chem.* **408**, 42 (1974).
212. H. Watelet, J.-P. Picard, G. Baud, J.-P. Besse & R. Chevalier, *Mater. Res. Bull.* **16**, 1131 (1981).
213. C. Dorémieux-Morin, J. P., Fraissard, J. P. Besse & R. Chevalier, *Solid State Ionics* **17**, 93 (1985).
214. U. Chowdhry, J. R. Barkley, A. D. English & A. W. Sleight, *Mater. Res. Bull.* **17**, 917 (1982).
215. D. J. Dzimitrowicz, J. B. Goodenough & P. J. Wiseman, *Mater. Res. Bull.* **17**, 971 (1982).
216. X. Turrillas, G. Delabouglise, J. G. Joubert, T. Fournier & J. Muller, *Solid State Ionics* **17**, 169 (1985).
217. B. D. Faithful & S. C. Wallwork, *Acta Crystallogr. B* **28**, 2301 (1972).
218. K. Kato & H. Saalfeld, *Acta Crystallogr. B* **33**, 1596 (1977).
219. P. Colomban, J.-P. Boilot, A. Kahn & G. Lucazeau, *Nouv. J. Chim.* **2**, 21 (1978).
220. W. A. England, A. J. Jacobson & B. C. Tofield, *J. Chem. Soc. Chem. Commun.* 895 (1976).
221. P. Colomban, G. Lucazeau, R. Mercier & A. Novak, *J. Chem. Phys.* **67**, 5244 (1977).
222. P. Colomban & A. Novak, *Solid State Commun.* **32**, 467 (1979).
223. J. C. Lassegues, M. Fouassier, N. Baffier, Ph. Colomban & A. J. Dianoux, *J. Physique* **41**, 273 (1980).
224. Ph. Colomban & G. Lucazeau, *J. Chem. Phys.* **72**, 1213 (1980).
225. W. Hayes, L. Holden & B. C. Tofield, *J. Phys. C.* **13**, 4217 (1980).
226. P. Colomban, G. Lucazeau & A. Novak, *J. Phys. C.* **14**, 4325 (1981).
227. A. R. Ochadlick, Jr, *Diss. Abstr. Internat.* **41**, 1819B (1980).
228. G. C. Farrington & J. L. Briant, *Mater. Res. Bull.* **13**, 763 (1978).
229. G. C. Farrington, J. L. Briant, M. W. Breiter & W. L. Roth, *J. Solid State Chem.* **24**, 311 (1978).
230. R. Haser & M. Pierrot, *Acta Crystallogr. B* **28**, 2538 (1972).
231. R. Haser & M. Pierrot, *Acta Crystallogr. B* **28**, 2542 (1972).
232. R. Haser, C. Penel & M. Pierrot, *Acta Crystallogr. B* **28**, 2548 (1972).
233. F. Payan & R. Haser, *Mater. Res. Bull.* **11**, 567 (1976).

234. H. Bode & G. Teufer, *Acta Crystallogr.* **8**, 611 (1955).
235. D. W. Davidson, L. D. Calvert, F. Lee & J. A. Ripmeester, *Inorg. Chem.* **20**, 2013 (1981).
236. D. W. Davidson & S. K. Garg, *Can. J. Chem.* **50**, 3515 (1972).
237. R. M. Izatt, B. L. Haymore & J. J. Christensen, *J. Chem. Soc. Chem. Commun.* 1308 (1972).
238. (a) B. L. Haymore, mentioned by I. Goldberg in *Inclusion Compounds, Vol. II* (J. L. Atwood, J. E. D. Davies & D. D. MacNicol, eds), Academic Press, New York, 1984, p. 303 (b) Yu. A. Simonov, N. F. Krasnova, A. A. Dvorkin, V. V. Yakshin, V. M. Abashkin & B. N. Laskorin, *Sov. Phys.-Dokl.* (*USA*) **28**, 823 (1983).
239. J.-P. Behr, P. Dumas & D. Moras, *J. Am. Chem. Soc.* **104**, 4540 (1982).
240. R. Chenevert, A. Rodrigue, M. Pigeon-Gosselin & R. Savoie, *Can. J. Chem.* **60**, 853 (1982).
241. G. S. Heo & R. A. Bartsch, *J. Org. Chem.* **47**, 3557 (1982).
242. R. Chenevert, A. Rodrigue, P. Beauchesne & R. Savoie, *Can. J. Chem.* **62**, 2293 (1984).
243. N. V. Shishkin, *Zap. Vses. Mineral. Obsch.* **79**, 94 (1950).
244. N. V. Shishkin, *Zh. Obsch. Khim.* **21**, 456 (1951).
245. J. Kubisz, *Bull. Acad. Pol. Sci., Geol. Geog.* **9**, 195 (1961).
246. G. P. Brophy & M. F. Sheridan, *Am. Mineral.* **50**, 1595 (1965).
247. N. N. Poprukailo & T. B. Shkodina, *Viniti* 1440, (1976); *Chem. Abstr.* **88** 123846e (1978).
248. R. W. T. Wilkins, A. Mateen & G. W. West, *Am. Mineral.* **59**, 811 (1974).
249. J. A. Ripmeester, C. I. Ratcliffe, J. E. Dutrizac & J. L. Jambor, *Can. Mineral* (1986), in press.
250. R. L. Parker, *Am. Mineral.* **47**, 127 (1962).
251. I. V. Ginzburg & G. V. Yukhnevich, *Geochem.* **31** (1962).
252. J. L. White & A. F. Burns, *Science* **141**, 800 (1963).
253. G. Brown & K. Norrish, *Mineral. Magn.* **29**, 929 (1952).
254. G. B. Bokii & D. K. Arkhipenko, *J. Struct. Chem.* **3**, 670 (1962).
255. D. K. Arkhipenko, G. B. Bokii & N. A. Palchik, *Geokhim. Mineral.* **98** (1980).
256. G. B. Bokii & D. K. Arkhipenko, *Phys. Chem. Minerals* **1**, 233 (1977).
257. Y. I. Tarasevich & F. D. Ovcharenko, *Dokl. Akad. Nauk SSSR* **161**, 1138 (1965).
258. V. V. Gordienko & G. E. Kalenchuk, *Zap. Vses. Mineral. Obshch.* **95**, 169 (1966).
259. K. Mereiter, *Tschermaks Min. Petr. Mitt.* **21**, 216 (1974).
260. D. K. Smith Jr, J. W. Gruner & W. N. Lipscomb, *Am. Mineral.* **42**, 594 (1957).
261. M. Ross & H. T. Evans Jr, *Amer. Mineral.* **49**, 1578 (1964).
262. M. Ross & H. T. Evans Jr, *Am. Mineral.* **50**, 1 (1965).
263. R. Sobry & M. Rinne, *J. Magn. Reson.* **12**, 152 (1973).
264. R. Sobry, *Am. Mineral.* **56**, 1065 (1971).
265. H. A. Szymanski, D. N. Stamires & G. R. Lynch, *J. Opt. Soc. Am.* **50**, 1323 (1960).
266. A. Corma, A. L. Agudo & V. Fornés, *J. Chem. Soc. Chem. Commun.* 942 (1983).
267. D. Mootz & D. Boenigk, *Z. Naturforsch. B* **39**, 298 (1984).

The Nature of the Hydrated Proton 203

268. D. Mootz, J. Goldmann & H. Wunderlich, *Angew. Chem. Int. Ed. Engl.* 8, 136 (1969).
269. D. Mootz & H. Wunderlich, *Z. Kristallogr.* 128, 447 (1969).
270. D. Mootz & H. Wunderlich, *Acta Crystallogr.* B 26, 1820 (1970).
271. A. Blaschette & H. Bürger, *Inorg. Nucl. Chem. Lett.* 3, 339 (1967).
272. J.-O. Lundgren & P. Lundin, *Acta Crystallogr.* B 28, 486 (1972).
273. J. Rozière & J. M. Williams, *J. Chem. Phys.* 68, 2896 (1978).
274. A. Eriksson & J. Lindgren, *Acta Chem. Scand.* 26, 1591 (1972).
275. J.-O. Lundgren, *Acta Crystallogr.* B 28, 475 (1972).
276. J.-O. Lundgren, *Acta Crystallogr.* B 35, 780 (1979).
277. D. D. Dexter, *Z. Kristallogr.* 134, 350 (1971).
278. S. K. Arora & Sundaralingam, *Acta Crystallogr.* B 27, 1293 (1971).
279. J.-O. Lundgren & J. M. Williams, *J. Chem. Phys.* 58, 788 (1973).
280. J. E. Finholt & J. M. Williams, *J. Chem. Phys.* 59, 5114 (1973).
281. L. J. Basile, P. LaBonville, J. R. Ferraro & J. M. Williams, *J. Chem. Phys.* 60, 1981 (1974).
282. C. I. Ratcliffe & B. A. Dunell, *J. Chem. Soc. Faraday Trans. 2*, 77, 2181 (1981).
283. R. Attig & D. Mootz, *Acta Crystallogr.* B 32, 435 (1976).
284. R. Attig & J. M. Williams, *Inorg. Chem.* 15, 3057 (1976).
285. R. Attig, *Cryst. Struct. Commun.* 5, 223 (1976).
286. R. Attig & D. Mootz, *Acta Crystallogr.* B 33, 2422 (1977).
287. R. Attig, *Acta Crystallogr.* A 31, S167 (1975).
288. R. Attig & J. M. Williams, *J. Chem. Phys.* 66, 1389 (1977).
289. D. Mootz & J. Fayos, *Acta Crystallogr.* B 26, 2046 (1970).
290. J. M. Williams, S. W. Peterson & H. A. Levy, in *Abstracts of the American Crystallographic Association Meeting, Albuquerque, New Mexico (1972), Vol. 17*, p. 51.
291. T. Gustafsson, *Acta Crystallogr.* C 41, 443 (1985).
292. N. K. Vyas, T. D. Sakore & A. B. Biswas, *Acta Crystallogr.* B 34, 3486 (1978).
293. N. K. Vyas, T. D. Sakore & A. B. Biswas, *Acta Crystallogr.* B 34, 2892 (1978).
294. J.-O. Lundgren, *Acta Crystallogr.* B 28, 1684 (1972).
295. J.-O. Lundgren & R. Tellgren, *Acta Crystallogr.* B 30, 1937 (1974).
296. B. M. Craven, S. Martinez-Carrera & G. A. Jeffrey, *Acta Crystallogr.* 17, 891 (1964).
297. E. K. Anderson, *Acta Crystallogr.* 22, 204 (1967).
298. E. K. Anderson & I. G. K. Anderson, *Acta Crystallogr.* B 31, 379 (1975).
299. J. M. Williams & S. W. Peterson, *Acta Crystallogr.* A 25, S113 (1969).
300. O. Simonsen & J. P. Jacobsen, *Acta Crystallogr.* B 33, 3045 (1977).
301. C.-O. Selenius & J.-O. Lundgren, *Acta Crystallogr.* B 36, 3172 (1980).
302. F. K. Winkler & J. D. Dunitz, *Acta Crystallogr.* B 31, 264 (1975).
303. R. A. Bell, G. G. Cristoph, F. R. Fronczek & R. E. Marsh, *Science* 190, 151 (1975).
304. J. E. Derry & T. A. Hamor, *J. Chem. Soc. Chem. Commun.* 1284 (1970).
305. H. J. van Wyk, *Z. Anorg. Allg. Chem.* 48, 1 (1906).
306. G. Mascherpa, *Rev. Chim. Mineral.* 2, 379 (1965).
307. L. H. Brickwedde, *J. Res. Nat. Bur. Stand.* 42, 309 (1949).
308. G. H. Cady & J. H. Hildebrand, *J. Am. Chem. Soc.* 52, 3834 (1930).
309. F. F. Rupert, *J. Am. Chem. Soc.* 31, 851 (1909).

310. G. Vuillard, C.R. Acad. Sci., Paris 241, 1308 (1955).
311. Gmelin Handbuch Anorg. Chem. 7, 214 (1931).
312. W. F. Giauque, E. W. Hornung, J. E. Kunzler & T. R. Rubin, J. Am. Chem. Soc. 82, 62 (1960).
313. R. G. Delaplane & J. A. Ibers, Acta Crystallogr. B 25, 2423 (1969).
314. D. Mootz & J. Goldmann, Z. Anorg. Allg. Chem. 368, 231 (1969).
315. S. Grimvall, Acta Chem. Scand. 25, 3213 (1971).
316. F. A. Cotton & G. Wilkinson, Advanced Inorganic Chemistry, Wiley, New York, fourth edn, 1980, p. 299.
317. D. Mootz & M. Steffen, Acta Crystallogr. B 37, 1110 (1981).
318. W. Joswig, H. Fuess & G. Ferraris, Acta Crystallogr. B 38, 2798 (1982).
319. M. Ishikawa, Acta Crystallogr. B 34, 2074 (1978).
320. G. A. Jeffrey, in Inclusion Compounds, Vol. I (J. L. Atwood, J. E. D. Davies & D. D. MacNicol, eds.), Academic Press, London, 1984, chap. 5, p. 148.
321. (a) W. F. Kuhs & M. S. Lehmann, J. Phys. Chem. 87, 4312 (1983); (b) W. F. Kuhs & M. S. Lehmann, this volume, p. 1.
322. R. Wang, W. F. Bradley & H. Steinfink, Acta Crystallogr. 18, 249 (1965).
323. S. Menchetti & C. Sabelli, Neus Jahrb. Mineral. Montasht. 406 (1976).
324. A. Potier, J. M. Leclercq & M. Allavena, J. Phys. Chem. 88, 1125 (1984).
325. K. Nakamoto, Infrared Spectra of Inorganic and Coordination Compounds, Wiley-Interscience, New York, 2 edn, 1970, p. 91.
326. D. Hadzi & S. Bratos, in The Hydrogen Bond, Vol. II (P. Schuster, G. Zundel & C. Sandorfy, eds), North-Holland, New York, 1976, chap. 12.
327. A. Novak, Structure and Bonding, 18, 177 (1974).
328. M. Falk & O. Knop, in Water: A Comprehensive Treatise, Vol. 2 (F. Franks, ed.), Plenum Press, New York, 1972, chap. 4.
329. M. E. Colvin, G. P. Raine, H. F. Schaefer III & M. Dupuis, J. Chem. Phys. 79, 1551 (1983).
330. B. S. Ault & G. C. Pimentel, J. Phys. Chem. 77, 57 (1973).
331. G. P. Ayers & A. D. E. Pullin, Spectrochim. Acta A 32, 1641 (1975).
332. A. Schriver, B. Silvi, D. Maillard & J. P. Perchard, J. Phys. Chem. 81, 2095 (1977).
333. L. Andrews & G. L. Johnson, J. Chem. Phys. 79, 3670 (1983).
334. G. Ritzhaupt & J. P. Devlin, J. Phys. Chem. 81, 521 (1977).
335. H. Kanno & J. Hiraishi, Chem. Phys. Lett. 107, 438 (1984).
336. H. Kanno & J. Hiraishi, Proc. IXth Int. Conf. Raman Spectroscopy, Tokyo, 1984, Tokyo, Chemical Society of Japan, 1984, p. 618.
337. D. W. Davidson & J. A. Ripmeester, in Inclusion Compounds, Vol. 3 (J. L. Atwood, J. E. D. Davies & D. D. MacNicol, eds), Academic Press, London, 1984, chap. 3.
338. N. Bloembergen, E. M. Purcell & R. V. Pound, Phys. Rev. 73, 679 (1948).
339. R. Kubo & K. Tomita, J. Phys. Soc. Japan 9, 888 (1954).
340. D. E. Woessner, J. Chem. Phys. 36, 1 (1962).
341. R. G. Barnes, Adv. Nucl. Quad. Reson. 1, 335 (1974).
342. H. Kiriyama & O. Nakamura, Bull. Chem. Soc. Japan 53, 635 (1980).
343. V. Gold, J. L. Grant & K. P. Morris, J. Chem. Soc. Chem. Commun. 397 (1976).
344. L. Glasser, Chem. Rev. 75, 21 (1975).
345. S. Lengyel & B. E. Conway, in Comprehensive Treatise of Electrochemistry,

The Nature of the Hydrated Proton   205

Vol. 5 (B. E. Conway, J. O'M. Bockris & E. Yeager eds), Plenum, New York, 1983, p. 365.
346. J. O'M. Bockris & A. K. N. Reddy in *Modern Electrochemistry*, *Vol. I*. Macdonald, London, 1970, chap. 5.
347. K.-D. Kreuer, A. Rabenau & W. Weppner, *Angew. Chem. Int. Ed. Engl.* 21, 208 (1982).
348. H. Gränicher, *Z. Kristallogr.* 110, 432 (1958).
349. M. Eigen & L. De Maeyer, *Proc. R. Soc., London, A* 247, 505 (1958).
350. W. B. Collier, G. Ritzhaupt & J. P. Devlin, *J. Phys. Chem.* 88, 363 (1984).
351. M. Kunst & J. W. Warman, *J. Phys. Chem.* 87, 4093 (1983).
352. E. Pines & D. Huppert, *Chem. Phys. Lett.* 116, 295 (1985).
353. F. Franks, in *Water: A Comprehensive Treatise*, *Vol. I*, (F. Franks, ed.), Plenum, New York, 1972, chap. 4, p. 141.
354. N. Bjerrum, *Science* 115, 385 (1952).
355. H. H. Richardson, P. J. Wooldridge & J. P. Devlin, *J. Phys. Chem.* 89, 3552 (1985).
356. J. Almlöf & U. Wahlgren, Theor. Chim. Acta 28, 161 (1973).
357. M. Fournier, M. Allavena & A. Potier, *Theor. Chim. Acta* 42, 145 (1976).
358. E. R. Andrew, in *Nuclear Magnetic Resonance*, Cambridge University Press, 1958, p. 156.
359. M.-H. Herzog-Cance, C. Belin & J.-F. Herzog, *C.R. Seances Acad. Sci., Ser. II* 298, 531 (1984).
360. R. Mercier, J. Douglade, P. G. Jones & G. M. Sheldrick, *Acta Crystallogr. C* 39, 145 (1983).
361. V. K. Trunov, V. A. Efremov, L. I. Konstantinova, Y. A. Velikodnyi & A. F. Golota, *Sov. Phys.-Dokl. (USA)* 25, 508 (1980).
362. S. C. Abrahams & J. L. Bernstein, *J. Chem. Phys.* 69, 4234 (1978).
363. R. Nedjar, M. M. Borel & B. Raveau, *Mater. Res. Bull.* 20, 1291 (1985).
364. P. Kebarle, in *Ions and Ion Pairs in Organic Reactions* (M. Szwarc, ed.), Wiley-Interscience, New York, 1972, chap. 2.
365. P. Kebarle, in *Ion Molecule Reactions* (J. L. Franklin, ed.), Plenum, New York, 1972, chap. 7.
366. P. Kebarle, in *Modern Aspects of Electrochemistry*, *No. 9* (B. E. Conway & J. O'M. Bockris, eds), Plenum, New York, 1974, p. 1.
367. R. W. Taft, in *Proton Transfer Reactions* (E. Caldin & V. Gold, eds), Chapman & Hall, London, 1975, p. 31.
368. P. Kebarle, *Annu. Rev. Phys. Chem.* 28, 445 (1977).
369. P. Kebarle & E. W. Godbole, *J. Chem. Phys.* 39, 1131 (1963).
370. P. Kebarle, S. K. Searles, A. Zolla, J. Scarborough & M. Arshadi, *J. Am. Chem. Soc.* 89, 6393 (1967).
371. A. Good, D. A. Durden & P. Kebarle, *J. Chem. Phys.* 52, 212 (1970).
372. D. P. Beggs & F. H. Field, *J. Am. Chem. Soc.* 93, 1567 (1971).
373. F. H. Field & D. P. Beggs, *J. Am. Chem. Soc.* 93, 1576 (1971).
374. S. L. Bennett & F. H. Field, *J. Am. Chem. Soc.* 94, 5186 (1972).
375. A. J. Cunningham, J. D. Payzant & P. Kebarle, *J. Am. Chem. Soc.* 94, 7627 (1972).
376. C. E. Young & W. E. Falconer, *J. Chem. Phys.* 57, 918 (1972).
377. Y. K. Lau, S. Ikuta & P. Kebarle, *J. Am. Chem. Soc.* 104, 1462 (1982).

206   C. I. Ratcliffe and D. E. Irish

378. M. DePaz, J. J. Leventhal & L. Friedman, *J. Chem. Phys.* **51**, 3748 (1969).
379. M. D. Newton & S. Ehrenson, *J. Am. Chem. Soc.* **93**, 4971 (1971).
380. M. Meot-Ner & F. H. Field, *J. Am. Chem. Soc.* **99**, 998 (1977).
381. G. G. Meisels, G. J. Sroka & R. K. Mitchum, *J. Am. Chem. Soc.* **96**, 5045 (1974).
382. G. G. Meisels, A. J. Illies, R. S. Stradling & K. R. Jennings, *J. Chem. Phys.* **68**, 866 (1978).
383. J. D. Payzant, A. J. Cunningham & P. Kebarle, *Can. J. Chem.* **51**, 3242 (1973).
384. S. G. Lias, J. F. Liebman & R. D. Levin, *J. Phys. Chem. Ref. Data* **13**, 695 (1984).
385. C. Y. Ng, D. J. Trevor, P. W. Tiedemann, S. T. Ceyer, P. L. Kronebusch, B. H. Mahan & Y. T. Lee, *J. Chem. Phys.* **67**, 4235 (1977).
386. J. E. Del Bene, H. D. Mettee, M. J. Frisch, B. T. Luke & J. A. Pople, *J. Phys. Chem.* **87**, 3279 (1983).
387. S. M. Collyer & T. B. McMahon, *J. Phys. Chem.* **87**, 909 (1983).
388. T. B. McMahon & P. Kebarle, *J. Am. Chem. Soc.* **107**, 2612 (1985).
389. S. S. Lin, *Rev. Sci. Instrum.* **44**, 516 (1973).
390. J. Q. Searcy & J. B. Fenn, *J. Chem. Phys.* **61**, 5282 (1974).
391. L. Kassner, Jr & D. E. Hagen, *J. Chem. Phys.* **64**, 1860 (1976).
392. G. M. Lancaster, F. Honda, Y. Fukuda & J. W. Rabalais, *Int. J. Mass Spectrom. Ion Phys.* **29**, 199 (1979); *J. Am. Chem. Soc.* **101**, 1951 (1979).
393. H. R. Carlon & C. S. Harden, *Appl. Opt.* **19**, 1776 (1980).
394. H. Udseth, H. Zmora, R. J. Beuhler & L. Friedman, *J. Phys. Chem.* **86**, 612 (1982).
395. D. Dreyfuss & H. Y. Wachman, *J. Chem. Phys.* **76**, 2031 (1982).
396. R. J. Beuhler & L. Friedman, *J. Chem. Phys.* **77**, 2549 (1982).
397. V. Hermann, B. D. Kay & A. W. Castleman, Jr, *Chem. Phys.* **72**, 185 (1982).
398. A. J. Stace and C. Moore, *Chem. Phys. Lett.* **96**, 80 (1983).
399. O. Echt, D. Kreisle, M. Knapp & E. Recknagel, *Chem. Phys. Lett.* **108**, 401 (1984).
400. H. Shinohara, U. Nagashima & N. Nishi, *Chem. Phys. Lett.* **111**, 511 (1984).
401. H. Shinohara, U. Nagashima, H. Tanaka & N. Nishi, *J. Chem. Phys.* **83**, 4183 (1985).
402. U. Nagashima, H. Shinohara, N. Nishi & H. Tanaka, *J. Chem. Phys.* **84**, 209 (1986).
403. E. Herbst & W. Klemperer, *Astrophys. J.* **185**, 505 (1973).
404. C. M. Leong, E. Herbst & W. F. Huebner, *Astrophys. J. Suppl. Ser.* **56**, 231 (1984).
405. T. de Jong, A. Dalgarno & W. Boland, *Astron. Astrophys.* **91**, 68 (1980).
406. D. Smith & N. G. Adams, *Int. Rev. Phys. Chem.* **1**, 271 (1981).
407. R. S. Narcisi & A. D. Bailey, *J. Geophys. Res.* **70**, 3687 (1965).
408. H. A. Schwarz, *J. Chem. Phys.* **67**, 5525 (1977).
409. M. D. Newton, *J. Chem. Phys.* **67**, 5535 (1977).
410. M. H. Begemann, C. S. Gudeman, J. Pfaff & R. J. Saykally, *Phys. Rev. Lett.* **51**, 554 (1983).
411. P. R. Bunker, T. Amano & V. Špirko, *J. Mol. Spectrosc.* **107**, 208 (1984).
412. N. N. Haese & T. Oka, *J. Chem. Phys.* **80**, 572 (1984).
413. P. Botschwina, P. Rosmus & E.-A. Reinsch, *Chem. Phys. Lett.* **102**, 299 (1983).

414. D.-J. Liu, N. Haese & T. Oka, *J. Chem. Phys.* **82**, 5368 (1985).

415. D.-J. Liu & T. Oka, *Phys. Rev. Lett.* **54**, 1787 (1985).

416. P. B. Davies, P. A. Hamilton & S. A. Johnson, *J. Opt. Soc. Am. B* **2**, 794 (1985).

417. B. Lemoine & J. L. Destombes, *Chem. Phys. Lett.* **111**, 284 (1984).

418. T. J. Sears, P. R. Bunker, P. B. Davies, S. A. Johnson & V. Špirko, *J. Chem. Phys.* **83**, 2676 (1985).

419. V. Špirko & P. R. Bunker, *J. Mol. Spectrosc.* **95**, 226, 381 (1982); V. Špirko, *J. Mol. Spectrosc.* **101**, 30 (1983).

420. G. M. Plummer, E. Herbst & F. C. DeLucia, *J. Chem. Phys.* **83**, 1428 (1985).

421. M. Bogey, C. Demuynck, M. Denis & J. L. Destombes, *Astron. Astrophys.* **148**, L11 (1985).

422. $T_{1\rho}$ and T_{1D} refer to spin–lattice relaxation times in the rotating frame and the dipolar field respectively. These are sensitive to motions of lower frequency compared to T_1. D. C. Ailion, *Adv. Magn. Reson.* **5**, 177 (1971).

Appendix: Studies of the hydrated proton in solids

Compound	Claimed species	References				
		X-ray structure	Neutron structure	Vibrational spectra	NMR	Other
Inorganic acids						
$HF \cdot H_2O$	H_3O^+	38	—	39, 40	—	—
$(HF)_2 \cdot H_2O$	H_3O^+	38, 41	—	—	—	—
$(HF)_4 \cdot H_2O$	H_3O^+	41	—	—	—	—
$HCl \cdot H_2O$	H_3O^+	42	—	39, 40, 43–47	—	—
$HCl \cdot 2H_2O$	$H_5O_2^+$	48	—	32, 44, 45, 47, 49	—	—
$HCl \cdot 3H_2O$	$H_5O_2^+$, $H_7O_3^+$	50	—	44, 45	—	—
$HCl \cdot 4H_2O$	$H_9O_4^+$, $H_7O_3^+$	—	—	44, 45	—	—
$HCl \cdot 6H_2O$	$H_9O_4^+$	51	—	—	—	—
$HBr \cdot H_2O$	H_3O^+	52	—	39, 40, 43–46	—	—
$HBr \cdot 2H_2O$	$H_5O_2^+$	52	53	32, 44, 45	54	—
$HBr \cdot 3H_2O$	$H_5O_2^+$	52	—	44, 45	—	—
$HBr \cdot 4H_2O$	$H_9O_4^+$, $H_7O_3^+$	55	—	44, 45, 56	—	—
$HI \cdot H_2O$	H_3O^+	—	—	39	—	—
$(CsCl)_4 \cdot 3HCl \cdot 3H_2O$	H_3O^+	58, 59	—	57	—	—
$(CsCl)_3 \cdot 2HCl \cdot H_2O$	H_3O^+	58, 59	—	—	—	—
$(CsBr)_3 \cdot 2HBr \cdot H_2O$	H_3O^+	—	—	—	—	—
$HClO_4 \cdot H_2O$	H_3O^+	15, 40, 60–62	63, 64	32, 40, 65–78	16, 17, 79–88	64, 89–94
$HClO_4 \cdot 2H_2O$	$H_5O_2^+$	95	—	32, 78, 96, 97	54	—
$HClO_4 \cdot 2\tfrac{1}{4}H_2O$	H_3O^+	98	—	—	—	—
$HClO_4 \cdot 3H_2O$	$H_7O_3^+$	99	—	100	101	—
$HClO_4 \cdot 3\tfrac{1}{2}H_2O$	$H_9O_4^+$	102	—	—	—	—
$(HNO_3)_4 \cdot H_2O$	H_3O^+	—	—	103	—	—
$HNO_3 \cdot H_2O$	H_3O^+	104, 105	—	32, 40, 65, 66, 69, 106–108	83, 86, 87	—

$HNO_3 . 3H_2O$	$H_7O_3^+$	109–111	—	100	101	—
$H_2SO_4 . H_2O$	H_3O^+	112–113	—	66, 69, 114	16, 87, 115, 116	—
$H_2SO_4 . 2H_2O$	H_3O^+	117	—	66, 114	115, 118	—
$H_2SO_4 . 4H_2O$	$H_5O_2^+$	119	—	114	—	—
$H_2SO_4 . 6\frac{1}{2}H_2O$	$H_7O_3^+$, $H_5O_2^+$	120	—	—	—	—
$H_2SO_4 . 8H_2O$	$H_5O_2^+$	120	—	—	—	—
$H_2SeO_4 . H_2O$	H_3O^+	121	—	118	—	—
$H_2SeBr_6 . 8H_2O$	$H_5O_2^+$	122	—	—	—	—
$H_2TeCl_6 . 2H_2O$	H_3O^+(?)	—	—	—	—	123
$H_2TeBr_6 . 2H_2O$	H_3O^+(?)	—	—	—	—	123
$(CF_3SO_3H)_2 . H_2O$	H_3O^+	124	—	—	—	—
$CF_3SO_3H . H_2O$	H_3O^+	125	126	—	88	—
$CF_3SO_3H . 2H_2O$	$H_5O_2^+$	127	—	—	88	—
$CF_3SO_3H . 4H_2O$	$H_9O_4^+$	128	—	—	—	—
$CF_3SO_3H . 5H_2O$	H_3O^+	129	—	—	—	—
$H_4P_2O_6 . 2H_2O$	H_3O^+	130	—	—	—	131
$(PO(OH)NH)_3 . 2H_2O$	$H_5O_2^+$	132	—	—	—	—
$(PO(OH)NH)_4 . 2H_2O$	H_3O^+	133–135	—	—	—	—
$ErH(C_2O_4)_2 . 3H_2O$	$H_5O_2^+$	136–137	—	—	—	—
$YH(C_2O_4)_2 . 3H_2O$	$H_5O_2^+$	137	138	139	84	—
$(PW_{12}O_{40}H_3) . 6H_2O$	$H_5O_2^+$	140	140	—	84	—
$HCr(SO_4)_2 . 7H_2O$	H_3O^+	141	—	—	—	—
$H_2Cr_4(SO_4)_7 . 24H_2O$	H_3O^+	142	—	—	—	—
$HIn(SO_4)_2 . 4H_2O$	$H_5O_2^+$	143	—	—	—	—
$HIn(SO_4)_2 . 2H_2O$	H_3O^+	—	—	144	144	144
$HTl(SO_4)_2 . 2H_2O$	H_3O^+	—	—	144	144	144
$HV(CF_3SO_3)_4 . 8H_2O$	$H_5O_2^+$	145	145	—	—	—
$HGa_3(SO_4)_2(OH)_6 . H_2O$	H_3O^+	146	—	—	147, 148	—
$InPO_4 . 2H_2O$	H_3O^+	149	—	—	—	—
$TlPO_4 . 2H_2O$	H_3O^+	149	—	—	—	—
$TlAsO_4 . 2H_2O$	H_3O^+	149	—	—	—	—
$HUO_2PO_4 . 4H_2O$ (HUP)	$H_5O_2^+$	150, 151	—	152–155	156–158	150, 154, 156, 159–163

Appendix *continued*

Compound	Claimed species	References				
		X-ray structure	Neutron structure	Vibrational spectra	NMR	Other
$HUO_2AsO_4 . 4H_2O$ (HUAs)	$H_5O_2^+$	150	164-166	167	156, 157 167	150, 156, 161, 167, 168
$H_2Sb_2(SO_4)_4 . 2H_2O$	H_3O^+	360	—	—	—	—
$HNb_3O_8 . H_2O$	H_3O^+	—	—	—	—	363
$(Co(en)_2Cl_2)Cl . HCl . 2H_2O$	$H_5O_2^+$	169	170-171	172	—	173
$(Co(en)_2Br_2)Br . HBr . 2H_2O$	$H_5O_2^+$	174	—	172	—	—
$(Cr(en)_2Cl_2)Cl . HCl . 2H_2O$	$H_5O_2^+$	175	—	172	—	—
$(Co(l-pn)_2Cl_2)Cl . HCl . 2H_2O$	$H_5O_2^+$	176, 177	—	172	—	173
$(Rh(en)_2Cl_2)Cl . HCl . 2H_2O$	$H_5O_2^+$	—	—	172	—	—
$(Rh(en)_2Br_2)Br . HBr . 2H_2O$	$H_5O_2^+$	—	—	172	—	—
$(Rh(py)_4Cl_2)Cl . HCl . 2H_2O$	$H_5O_2^+$	—	—	172	—	—
$(Rh(bipy)_2Cl_2)Cl . HCl . 2H_2O$	$H_5O_2^+$	—	—	172	—	—
$NaIO_4 . 3H_2O$	H_3O^+	362	—	—	178	—
$Na_2SiO_3 . 9H_2O$	H_3O^+	—	—	—	—	—
$H_2SiF_6 . 4H_2O$	$H_5O_2^+$	120, 179	—	—	—	—
$H_2SiF_6 . 6H_2O$	$H_5O_2^+$	120, 179	—	—	—	—
$H_2SiF_6 . 9\frac{1}{2}H_2O$	$H_7O_3^+, H_5O_2^+$	120	—	—	—	—
$HBF_4 . H_2O$	H_3O^+	180, 181	—	—	—	—
$HBF_4 . 2H_2O$	$H_5O_2^+$	180, 181	—	—	—	—
$HWOF_5 . H_2O$	H_3O^+	—	—	182	—	182
$HNbF_6 . H_2O$	H_3O^+	—	—	182	—	182
$HTaF_6 . H_2O$	H_3O^+	—	—	182	—	182
$HIrF_6 . H_2O$	H_3O^+	—	—	183	—	183
$HRuF_6 . H_2O$	H_3O^+	—	—	183	—	183
$HTiF_5 . H_2O$	H_3O^+	—	—	184	—	—
$HBiF_6 . H_2O$	H_3O^+	—	—	185	—	—

Compound	Ion					
$HUF_6 \cdot H_2O$	H_3O^+	—	—	186	—	—
$HPtF_6 \cdot H_2O$	H_3O^+	—	—	183	—	183
$H_2PtF_6 \cdot H_2O$	H_3O^+	—	—	183	—	183
$HAsF_6 \cdot H_2O$	H_3O^+	120, 187	187	187, 188	—	187
$HSbF_6 \cdot H_2O$	H_3O^+	187	187	187, 188	—	187
$HSbF_6 \cdot 2H_2O$	$H_5O_2^+$	—	—	189	—	189
$HSbF_5OH \cdot H_2O$	H_3O^+	—	—	189	—	189
$HPtCl_5 \cdot 2H_2O$	$H_5O_2^+$	—	—	190	83, 118	—
$H_2PtCl_6 \cdot 2H_2O$	H_3O^+	—	—	172, 190	—	—
$\cdot 4H_2O$	$H_5O_2^+$	—	—	190	—	—
$\cdot 6H_2O$	$H_7O_3^+$	—	—	190	—	—
$(H_3O)_{1.6}Pt(C_2O_4)_2 \cdot nH_2O$	H_3O^+	192	—	—	—	191
$HK_8Rh_3Br_{18} \cdot 10H_2O$	H_3O^+	—	—	193	—	193
$[HRh_2Cl_9][Pr_4N]_2 \cdot H_2O$	H_3O^+	—	—	193	—	193
$[HRh_2Br_9][Bu_4N]_2 \cdot H_2O$	H_3O^+	—	—	193	—	193
$H_2[(Mo_6Cl_8)Cl_6] \cdot 8H_2O$	$H_5O_2^+$	194	—	194	—	—
$H[Mo_2Cl_8H][MoCl_4O][Et_4N]_3 \cdot 3H_2O$	H_3O^+	195	—	—	—	—
$H_2(Mo_6Cl_{14}) \cdot 2H_2O$	$H_5O_2^+$	196–197	—	172	—	198
$HZn_2Cl_5 \cdot 2H_2O$	$H_5O_2^+$	198	199	—	—	—
$HICl_4 \cdot 4H_2O$	$H_5O_2^+$	—	—	—	—	—
$HAuCl_4 \cdot 4H_2O$	H_3O^+	201	—	—	200	—
$HAuCl_4 \cdot H_2O \cdot (C_9H_{10}N_2O_2)$	$H_5O_2^+$	202	—	203	—	203
$HAu(CN)_4 \cdot 2H_2O$	$H_5O_2^+$	—	—	32, 204	—	—
$HGaCl_4 \cdot 2H_2O$	$H_5O_2^+$	—	—	32, 204	—	—
$HGaBr_4 \cdot 2H_2O$	$H_5O_2^+$	206	—	46, 205	54	—
$HSbCl_6 \cdot H_2O$	H_3O^+	—	—	32, 204	—	—
$HSbCl_6 \cdot 2H_2O$	$H_5O_2^+$	209	—	207, 208	—	—
$HSbCl_6 \cdot 3H_2O$	$H_{14}O_6^{2+}$	—	—	207	—	—
$HSbCl_6 \cdot 4H_2O$	$H_9O_4^+$	210	—	207	—	—
$HSbCl_6 \cdot 4\tfrac{1}{2}H_2O$	$H_5O_2^+$	—	—	—	—	—
$HSbCl_5OH \cdot H_2O$	H_3O^+	—	—	211	—	—

Appendix *continued*

Compound	Claimed species	References X-ray structure	Neutron structure	Vibrational spectra	NMR	Other
$HSbCl_5OH . 2H_2O$	$H_5O_2^+$	—	—	211	—	—
$HSbCl_5OH . 3H_2O$	$H_7O_3^+$	—	—	211	—	—
$HSbO_3 . nH_2O$	H_3O^+	212	—	—	213	212, 214, 215
$HSbTeO_6 . nH_2O$	H_3O^+	217	—	—	—	216
$(C_6H_5)_4AsCl . HCl . 2H_2O$	$H_5O_2^+$	218, 219	—	—	—	—
Hydrated β-Alumina and isomorphous gallates	H_3O^+	—	220	221–226, 224	227	228, 229
$H_2(NO_2)_9(ClO_4)_{11} . 2H_2O$	H_3O^+	359	—	—	83	—
$H(NO_2)(ClO_4)_2 . H_2O$	H_3O^+	230	—	—	83	—
$H_{3.5}(Fe^{II}_{0.5} \, Fe^{III}(CN)_6)H_2O$	$H_5O_2^+$	231	—	—	—	—
$H_{3.6}(Fe^{II}_{0.6} \, Fe^{III}(CN)_6) . 1.6H_2O$	$[H_9O_4 . H . H_9O_4]^{2+}$	232	—	—	—	—
$H_{3.33}(Fe^{II}_{0.33} \, Fe^{III}(CN)_6) . 2.66 \, H_2O$	$H_5O_2^+$	233	—	—	—	—
$HK_2Fe(CN)_6 . H_2O$	H_3O^+	233	—	233	—	—
$HPF_6 . HF . 5H_2O$	H_3O^+	234, 235	—	—	235, 236	—
$HAsF_6 . HF . 5H_2O$	H_3O^+	235	—	—	235	—
$HSbF_6 . HF . 5H_2O$	H_3O^+	235	—	—	235	—
$HBF_4 . 5H_2O$ Str I	(H_3O^+)	120	—	—	—	—
$HClO_4 . 5\tfrac{1}{2}H_2O$ Str I	(H_3O^+)	120	—	—	—	—
$HPF_6 . 7.7H_2O$ Str I	(H_3O^+)	120	—	—	—	—
$HAsF_6 . 6H_2O$ Str IV	(H_3O^+)	120	—	—	—	—
$HSbF_6 . 6H_2O$ Str IV	(H_3O^+)	120	—	—	—	—
$H_2AlF_5 . 5H_2O$	H_3O^+	361	—	—	—	—
$H_3AlF_6 . 6H_2O$	$H_5O_2^+$	120	—	—	—	—
$H_2TiF_6 . 3H_2O$	H_3O^+, $H_5O_2^+$	120	—	—	—	—
$H_2TiF_6 . 6H_2O$	$H_5O_2^+$	120	—	—	—	—
$HClO_4 . $ dibenzo-18-crown-6 $. H_2O$	H_3O^+	—	—	237	—	—
$HClO_4 . $ dicyclohexyl-18-crown-6 $. H_2O$	H_3O^+	238	—	—	—	—
HCl-tetracarboxylic-18-crown-6 $. H_2O$	H_3O^+	239	—	—	—	—
$HBF_4 . $ 18-crown-6 $. H_2O$	H_3O^+	—	—	240	240	—
Oxonium crown ether complexes	H_3O^+	—	—	241, 242	88, 241	—

Compound	Species					
Jarosite, H_3O^+ replaces K^+ in $KFe_3(SO_4)_2(OH)_6$	H_3O^+	243–246	—	—	—	—
Alunite, H_3O^+ replaces K^+ in $KAl_3(SO_4)_2(OH)_6$	H_3O^+	247, 250	84, 248, 249	247	—	—
Amphiboles	H_3O^+	—	—	251	—	—
Micas	H_3O^+	253	—	252	—	—
Vermiculite	H_3O^+	254	—	254–256	—	—
Polygorskite	H_3O^+	257	—	—	—	—
Spodumene H_3O^+ replaces Li^+	H_3O^+	258	—	—	—	—
Rhomboclase $H_5O_2Fe(SO_4)_2 \cdot 2H_2O$	$H_5O_2^+$	—	—	—	—	259
Uranophane $(Ca(H_3O^+)_2)UO_2(SiO_4)_2 \cdot 3H_2O$	H_3O^+	—	—	—	—	260
Torbernites	$H_3O^+,\ H_5O_2^+$	262	248	248	—	261
Hydrated Uranates	H_3O^+	264	263	—	—	—
Zeolites	H_3O^+	—	—	265, 266	—	—
Organic acids						
$CF_3COOH \cdot 4H_2O$ (Trifluoro-acetic)	$H_5O_2^+$	—	—	—	—	267
$(CH_2SO_3H)_2 \cdot 2H_2O$ (Ethane-disulphonic)	H_3O^+	—	—	271	—	268–270
$C_6H_5SO_3H \cdot 3H_2O$ (Benzene-sulphonic)	$H_7O_3^+$	—	101	—	—	—
$C_6H_3Cl_2SO_3H \cdot 3H_2O$ (dichloro-benzene-sulphonic)	$H_7O_3^+$	—	101	274	273	272
$C_6H_3Br_2SO_3H \cdot 3H_2O$ (dibromo-benzene-sulphonic)	$H_7O_3^+$	—	—	274	276	275
$CH_3C_6H_4SO_3H \cdot H_2O$ (Para-toluene-sulphonic)	H_3O^+	—	88, 282	281	279, 280	277, 278
$C_6H_4(COOH)SO_3H \cdot 3H_2O$ (Ortho-sulphobenzoic)	$H_5O_2^+$	—	—	—	284	283
$C_6H_3(COOH)(OH)SO_3H \cdot 2H_2O$ (5-sulpho-salicylic)	$H_5O_2^+$	—	—	—	288	285–287
$C_6H_3(COOH)(OH)SO_3H \cdot 3H_2O$ (5-sulpho-salicylic)	$H_7O_3^+$	—	—	—	290	286, 287, 289

Appendix *continued*

Compound	Claimed species	References				
		X-ray structure	Neutron structure	Vibrational spectra	NMR	Other
$C_6H_3(COOH)(OH)SO_3H \cdot 4H_2O$ (5-sulpho-salicyclic)	$H_9O_4^+$	120	—	—	—	—
$CH_3C_6H_2(COOH)(OH)SO_3H \cdot 4H_2O$ (4-methyl-5-sulpho-salicylic)	$H_7O_3^+$, H_3O^+	291, 292	—	—	—	—
$CH_3C_6H_2(COOH)(OH)SO_3H \cdot 2H_2O$ (5-methyl-3-sulpho-salicylic)	$H_5O_2^+$	293	—	—	—	—
$C_6H_2(NO_2)_3SO_3H \cdot 4H_2O$ (Picryl-sulphonic)	$H_5O_2^+$	294	295	—	—	—
$C_4H_3O_5N_3 \cdot 3H_2O$ (Diituric)	H_3O^+ (?)	296	—	—	—	—
$C_6O_2(NO_2)_2(OH)_2 \cdot 6H_2O$ (Nitranilic)	$H_7O_3^+$	297, 298	299	—	—	—
$C_6O_2(CN)_2(OH)_2 \cdot 6H_2O$ (Cyananilic)	$H_7O_3^+$	298	—	—	—	—
$C_9H_4O_4NH \cdot 2H_2O$ (2-nitro-1,3-indandione)	$H_5O_2^+$	300	301	—	—	—
$(C_9H_{17}ON)_3 \cdot HCl \cdot H_2O$ (tripelargolactam Hydrochloric)	H_3O^+	302	—	—	—	—
$[C_9H_{18})_3(NH)_2Cl]^+$ $Cl^- \cdot HCl \cdot 6H_2O$	$H_{13}O_6^+$	303	—	—	—	—
$[(C_{14}H_{16}N_2)Br_2 \cdot H_2O]_3 \cdot HBr$ (1,1'-tetramethylene-2,2'-dipyridylium dibromide)	$H_7O_3^+$	304	—	—	—	—

(?) indicates possible H_3O^+ not suggested by original authors. (H_3O^+), these parentheses show that the hydrated proton forms part of a cage structure.